This fascinating and absorbing book by Kah Seng Loh and Li Yang Hsu is a classic in the social history of medicine. Drawing on a broad range of archival sources, including vivid patients' accounts, the authors use the history of tuberculosis control in Singapore as a way to highlight key themes in the city-state's social and political history, including the development of the state and the shifting lines of social and economic inequality. This book will be of interest to scholars of health and society around the world as a richly detailed case study that is sure to illuminate wider comparisons. As a collaboration between a social historian and a physician who specializes in infectious disease, it is also a model of interdisciplinary scholarship.

Sunil Amrith, *Harvard University*

Tuberculosis – The Singapore Experience, 1867–2018

Through a rich account of tuberculosis in Singapore from the mid-nineteenth century to the present day, this book charts the relationship between disease, society and the state, outlining the struggles of colonial and postcolonial governments to cope with widespread disease and to establish effective public health programmes and institutions. Beginning in the nineteenth century when British colonial administrators viewed tuberculosis as a racial problem linked to the poverty, housing and insanitary habits of the Chinese working class, the book goes on to examine the ambitious medical and urban improvement initiatives of the returning British colonial government after the Second World War. It then considers the continuation and growth of these schemes in the postcolonial period and explores the most recent developments, which include combating the resurgence of TB and the rise of antimicrobial resistance. Throughout, the book highlights the special difficulties of Singapore as an open port city with a large multicultural population, discusses the development of specific government and non-governmental institutions (especially the Singapore Anti-Tuberculosis Association), and describes people's varied experiences, responses and resistance to the disease.

Kah Seng Loh is an Honorary Research Fellow at the University of Western Australia

Li Yang Hsu is Head of the Infectious Diseases Programme at the Saw Swee Hock School of Public Health, National University of Singapore.

Routledge Studies in the Modern History of Asia

For a full list of available titles please visit: www.routledge.com/Routledge-Studies-in-the-Modern-History-of-Asia/book-series/MODHISTASIA

Tuberculosis – The Singapore Experience, 1867–2018

Disease, Society and the State

Kah Seng Loh and Li Yang Hsu

LONDON AND NEW YORK

First published 2020
by Routledge
2 Park Square, Milton Park, Abingdon, Oxon OX14 4RN

and by Routledge
52 Vanderbilt Avenue, New York, NY 10017

Routledge is an imprint of the Taylor & Francis Group, an informa business

First issued in paperback 2021

British Library Cataloguing-in-Publication Data
A catalogue record for this book is available from the British Library

Library of Congress Cataloging-in-Publication Data
A catalog record has been requested for this book

ISBN: 978-0-367-35453-4 (hbk)
ISBN: 978-1-03-208443-5 (pbk)
ISBN: 978-0-429-33144-2 (ebk)

Typeset in Times New Roman
by Wearset Ltd, Boldon, Tyne and Wear

Contents

Acknowledgements

Our book, written by a historian and a physician both from the small island city-state of Singapore in Southeast Asia, attempts to combine the disciplines of history and medicine. Forward-looking and successful in many ways, Singapore is nevertheless grappling with the difficult issue of tuberculosis like many countries around the world. Like these countries, Singapore also has a long and largely uncharted history of tuberculosis and tuberculosis control.

We would like to thank the Saw Swee Hock School of Public Health, National University of Singapore, for providing invaluable support and funding for the research that made this book possible through the Infectious Diseases Programme grant. The programme aims to improve the understanding of infectious diseases in Singapore and the region, and ultimately to mitigate their impact by conducting rigorous research that can be translated into public health policies and practices. The support for the book, amid the programme's current priorities, recognises the role of history towards achieving these goals.

In particular, we would like to express our appreciation and gratitude to the following persons from the school: Prof Kee Seng Chia, Prof Yik Ying Teo, Ms Ai Li Quake, Ms Po Jan Chen, Ms Zunairah binti Lukman, and Ms Sharon Lee. We also wish to thank the staff of the Singapore Tuberculosis Elimination Programme and Tuberculosis Control Unit, who continue to work towards the elimination of the threat of tuberculosis from Singapore.

Documenting the history of tuberculosis has taken us to the archives and more broadly to social memory. We are grateful to the National Archives of Singapore for facilitating our research, particularly Mr Eric Chin, Ms Fiona Tan, Ms Gayathri Kaur Gill, and Ms Abigail Huang, and to our research assistants, Ms Zihan Loo, Ms Vaani Parameshwari Kiran Chaudhari, Mr V.P. Vishnu Prasad, Ms Teresa Barre, Ms Siti Nurain, Ms Dafina Kajtazi, and Ms Valentina Jokic.

We learned much from conversations and discussions – academic and otherwise – with various people, including Dr Nicholas White, Dr Tony Webster, Dr Barry Doyle, Dr Alistair Martyn Chew, Prof Paul Anantharajah Tambyah, Dr David Allen, Dr Anita Lundberg, Mr Edmund Arozoo, Dr Heong Hong Por, Mr Yoong How Hsien, Ms Nur Sakinah Rahmat, Mr Harbhajan Singh, Ms Meeravathy, Dr Keng We Koh, Dr Kai Khiun Liew, Mr Joo Teng Teh, Mr Dan Feng Tan, Dr Guo-Quan Seng, Dr Geoffrey Pakiam, Mr Michael Yeo, and Mr Alex Tan.

We also wish to thank the publishers at Routledge for greenlighting a manuscript on the history of medicine in Singapore, particularly Mr Peter Sowden, Editor for Asia, Russia and Eastern Europe, and his publishing and editorial colleagues.

Finally, we are indebted to the individuals whom we interviewed for their memories and stories, namely Prof Chin Hin Chew, Dr Seng Kee Teo, Prof Kee Tai Goh, A/Prof Cynthia Chee, Mr Boon H. Tang, Ms Wendy Tan, Ms Pushparani, and Ms Chew Yin Leong.

Kah Seng and Li Yang
April 2019

Abbreviations

AIDS	acquired immunodeficiency syndrome
BCG	Bacille-Calmette-Guérin
BMRC	British Medical Research Council
DTBC	Department of Tuberculosis Control
GH	General Hospital
HDB	Housing and Development Board
KKH	Kandang Kerbau Hospital
MDR-TB	multidrug-resistant tuberculosis
PAP	People's Action Party
SATA	Singapore Anti-Tuberculosis Association
SGH	Singapore General Hospital
SIT	Singapore Improvement Trust
STEP	Singapore Tuberculosis Elimination Programme
TB	tuberculosis
TBCU	Tuberculosis Control Unit
TTSH	Tan Tock Seng Hospital
UNICEF	United Nations International Children's Emergency Fund
URA	Urban Redevelopment Authority
WHO	World Health Organization

Introduction

Tuberculosis is written into the history of Singapore as an open city-state. An ancient disease prevalent throughout the world, it was tied to the government, economy and society of the small island state of a mere 600 square kilometres in Southeast Asia from the second half of the nineteenth century. Both the prevalence and control of the disease highlighted Singapore's wide-ranging economic, social and medical connections with the region and the world, and locally, the typology of urban housing built for immigrant labour. After the Second World War, the joint efforts of the government and community to combat tuberculosis helped define the shape of the Singapore state and society across the late-colonial and postcolonial periods. Our book records the long history of tuberculosis control in Singapore, but it also charts the rich narrative of the city-state through illness, disease control and recovery.

The more manifestly pivotal years of tuberculosis control in Singapore, as in many countries, were the three decades immediately after the war. In this time, the disease received the greatest political and social attention and was brought under control by a defining group of medical and environmental reforms, including, but not only limited to, antibiotic treatment. This also coincided with the period when Singapore experienced a difficult process of decolonisation from British rule to become a sovereign state in 1965. Tuberculosis control took form as a concerted national programme led primarily by the state, contributing to the making of a robust nation and healthy citizens. Thus, after his People's Action Party (PAP) registered a decisive victory over leftist opposition in the 1963 elections, Prime Minister Lee Kuan Yew utilised effective tuberculosis control as a metaphor for model Singaporeans:

> Unite the people, build a prosperous and an equal society, isolate the Communists, contain them, like the tuberculosis bacilli is contained, where you throw a hard crust around an infected wound and keep your patient healthy and he would live a good life.[1]

Lee's speech was striking but not unique, for in the Australian context, tuberculosis control aimed not only to treat patients but also to govern healthy citizens.[2]

In Singapore, there was also an important strand of social history and memory in addition to the politicians' response to tuberculosis. For many people who were born during or after the war, the disease was also historic, albeit in a somewhat different way. Tuberculosis defined the difficult times in which they lived (or died), as Tay Chin Tay recounted her husband's demise from the disease in an oral history interview:

> When my husband was diagnosed with tuberculosis (TB), in the past, there was no cure for TB. It is not like now ... the moment one is diagnosed with TB, he would be cured if he seeks treatment. In those days, when one contracts TB, he would have to prepare a coffin. When my husband was down with TB, there was no cure for him then.[3]

Clearly, the oral history suggests, if the state had forgotten tuberculosis, people had not. The book shows how histories of policy and citizenship in Singapore were intertwined with varied social experiences, responses and memories.

A long history of tuberculosis

Besides the post-war developments, there was also a long and hitherto largely veiled history of tuberculosis in colonial Singapore, although this was one which existed mostly in the writings and debates of colonial officials and doctors rather than in actual policy or concrete achievements. It took more than 60 years from Robert Koch's discovery of the *Mycobacterium tuberculosis* in 1882 before a genuine anti-tuberculosis programme emerged in Singapore after the war in 1948. But as it was recognised as a mounting public health threat in the late nineteenth and early twentieth centuries, the disease became a social and intellectual construct and a political battlefield.

The period from 1867, when the government of Singapore passed to the Colonial Office in Britain, to 1941, when British rule was temporarily disrupted by the Japanese invasion of Southeast Asia, was a formative phase in Singapore's history. Sometimes simplistically dismissed as one of laissez-faire colonialism, the period was one of increasing direct rule as the British became more interested, concerned and (occasionally) committed to the island's social and economic affairs.[4] On the one hand, they discarded the long-standing practice of appointing Chinese *kapitans* ('captains', or community leaders) to directly operate the lucrative opium and other revenue farms themselves, while also appointing 'protectors' to govern the majority immigrant Chinese population.[5] The entrepôt trade of Singapore attracted numerous short-term male immigrants from rural China and India who worked as labourers of various types. In effect Singapore became a coolie town, with lowly paid migrants living en masse in congested and insanitary dwellings in the municipal area.[6] Tuberculosis appeared in the medico-official mind in such a context from the 1890s. It grew historically significant as officials and doctors imagined and debated the cleansing of the deathly quarter of what was otherwise a prosperous port, and the management of the ethnically diverse working class which comprised the requisite labour.

Colonial interest in tuberculosis as a disease of the Chinese thus added to the expanding direct colonial governance of Singapore society. One of the little-known effects of these early tuberculosis ruminations was to lay the basis for the remarkable public housing state which emerged in Singapore after the war. As a result, four-fifths of Singaporeans presently live in government-built flats, which most of them purchase on 99-year leases. At the dawn of the twentieth century, colonial officials and doctors had connected tuberculosis control to the housing habits of the coolies (unskilled labourers in the local parlance). In Singapore, as in other urban situations, the disease was not simply an illness to be treated in a hospital: where and how one lived and one's social relations were deemed to be equally crucial. In the municipal officials' damning assessment of the insanitary workers' housing in 1933, 'The Commissioners' byelaws provide better housing for pigs'.[7]

It was in this context that the British sanitary expert W.J.R. Simpson arrived to study Singapore's shophouses in 1906 and tuberculosis control provided a compelling template for the state to regulate and transform urban space. This greatly assisted the historic programmes of urban renewal and public housing development after the war. It played an important role in the social ascendancy and popularity of state-built flats in Singapore. There is at least one similar instance in global history: Hong Kong, another city-state in Asia where the threat of tuberculosis helped convert British disinterest in ruling the colony and its population. There, the disease was likewise surmised as a problem of the slums, becoming a major impetus for the post-war public housing programme.[8] But more so than in Hong Kong, tuberculosis control and public housing in Singapore have highlighted to the citizenry how the PAP government was single-minded in tackling socio-economic issues that affected the general population. They played major roles in generating popular support for the PAP since 1959.

While the municipal administrators of early twentieth-century Singapore considered ways to improve the urban environs, they also did much more to shape the society. Their proposals to sanitise the town were accompanied by charts on the breakdown of tuberculosis deaths by ethnicity and gender and, perhaps more tellingly, by commentary on the differing degrees of immunity and resistance among the diverse social groups which populated Singapore. The administrators fixed their gaze on the labouring class of one particular ethnic group: the Chinese. Tuberculosis thus helped to harden the divisions within the population of the city-state. Further, both European and Western-trained Chinese physicians began to apply the prevailing theory of racial immunity and resistance to the population. While tuberculosis was most prevalent among the Chinese, they were deemed to have greater resistance to it compared to the Malays and Indians. It was only after the 1950s that these racial discourses gave way to national interpretations, but racial categories continued to be used to classify tuberculosis cases (and the general population).[9]

The influence of race in Singapore resonates with important studies elsewhere, both historical and contemporary. It emerged from the evolutionary theory of tuberculosis in the early twentieth century, where 'primitive peoples' living in 'virgin soil' areas were deemed to lack the immunity and resistance which more 'advanced' races possessed.[10] In northern Norway in the first half of the twentieth century, the anti-tuberculosis campaign among the indigenous Sami people was framed as a war against the uncivilised, but this resulted in the displacement of their traditional housing and language.[11] In more recent times, racial interpretations have been made of immigrant groups, as in post-war Britain, where tuberculosis control precipitated or heightened moral panics against the Irish, Indians and Pakistanis.[12] In the US, the illness provoked nativist efforts to exclude both internal and international migrants in the nineteenth and twentieth centuries,[13] and in Australia, it was linked especially to the aboriginal people and Asian immigrants.[14] Racial-migrant accounts of tuberculosis were ways of demarcating and managing not only diseases, but also social and ethnic diversity. This is relevant for an open city-state like Singapore which, as elsewhere, continues to rely heavily on the recruitment of low-wage Asian migrant labour.

Reflexive upon success

It may seem that the narrative of tuberculosis control in Singapore is one of success. Once the late-colonial regime, the PAP administration and the community decided to make the disease their focal point after the war, Singapore made great strides in the 1950s and 1960s. Tuberculosis control underscored the leading health policy role of the PAP government, which has governed Singapore without interruption since the 1959 elections. More than the British, the PAP was able to implement and coordinate a nationwide campaign against the disease – no less than an 'action programme'. The mainstream perspective of tuberculosis control, which reads along the official grain to stress the role of the government and public hospitals, duly frames a linear narrative of accomplishment.[15] This heralds a different experience from places like India, where as Sunil Amrith observes, a similarly ambitious campaign was thwarted by numerous difficulties, with biomedicine failing to make much headway into Indian society.[16]

Likewise, in his rather pessimistic assessment of World Health Organization (WHO) and national tuberculosis control programmes in American, African and Asian contexts, Christian McMillen has found mostly failures, particularly the inability to learn from history. Singapore's post-war efforts were closely modelled upon and aided by the overarching programme of WHO and the United Nations International Children's Emergency Fund (UNICEF), which constituted the 'largest mass action the world has ever known against one single disease'.[17] Singapore also embarked on a successful policy of industrialisation in the 1960s and was arguably one of the few places in the world which could claim to have realised McMillen's description of the aims of the WHO-UNICEF anti-tuberculosis programme: 'Biomedicine was at the heart of the movement: formerly sick people made well could then become productive citizens'.[18] Singapore's immigrant coolies transitioned into disciplined industrial workers in this period of tuberculosis control and rapid industrialisation.

Singapore's contrasting and impactful experience with tuberculosis highlights several points. The first is fairly obvious: generalisations such as McMillen's, which are based on a collection of case studies, have to accommodate contrary instances. This is not limited to achievements either: for example, Singapore continued to subscribe to the Bacille-Calmette-Guérin (BCG) vaccination regimen even after growing evidence from India about its lack of efficacy against pulmonary tuberculosis. Second, it would not do to simply dismiss Singapore as an exception on account of its small size as a city-state. The island's openness to immigrants (and their microbes) made it doubly susceptible to infectious diseases, while its anti-tuberculosis programme was comparably ambitious and far-reaching like those in larger states and capital cities. Singapore succeeded primarily not because of its size but because of the political and social commitment to combating the disease, drawing upon a long period of debate and experimentation.

Third, just as striking as the apparent success was the deep reflexivity of the Singapore state. The island's accomplishments in tuberculosis control were not absolute: even after the 1960s, officials and doctors remained concerned with the disease among some social groups, particularly the elderly, as they were with patients not completing their treatment. Both issues prompted the government to launch the Singapore Tuberculosis Elimination Programme (STEP) in 1997. This parallels the Korean experience, where initial success with chemotherapy was followed by the resurgence of tuberculosis due to an ageing population and rising immigration (much greater in Singapore's case).[19] As McMillen points out, patient compliance with drug therapy had been low in countries like Kenya. But while he maps a frustrating global experience of discovering, forgetting and re-finding the disease, Singapore's case was qualitatively different – tuberculosis was never really off the record.[20] Although the city-state experienced a similar de-emphasis on tuberculosis relative to other diseases from the late 1960s, Singapore did not lose sight of the illness. State officials and physicians continued to worry over the disease's trends and demographic characteristics, as did the leaders of the single most important non-governmental organisation for the disease, the Singapore Anti-Tuberculosis Association (SATA, discussed below).

The reflexive Singapore state was quintessentially modernist: it possessed a sustained sense of anxiety which drove the political will to reform and improve. This reflexivity was not removed by results, which merely redoubled it. It was also characteristic of the city-state's national culture of vulnerability, which belied its apparent economic and social achievements. The state deems the small, accessible island it governs, which has few natural resources and whose only real resource is its people, to be perpetually susceptible to external and internal threats.[21] Arguably, the threat of tuberculosis infection has played a crucial role in exemplifying and enhancing the sense of vulnerability in Singapore up to the present day.

Social history: patients and SATA

The history of tuberculosis in Singapore would be incomplete without its social history. As much as the state took the lead against the disease, its actual role and impact depended on two important social actors: the patients (and their families) and the community. While Singapore was a small city-state, governing its diverse immigrant population had proven difficult for the colonial government and to a lesser extent even the PAP. The Chinese working class historically preferred to live in shophouse cubicles which offered cheap rents for single men. After the war, larger families migrated to semi-autonomous and unauthorised settlements made of wood and thatch, called kampongs (villages in Malay). The British and PAP labelled such mass housing as 'slums' and 'squatter areas', but their attempts to remove or regulate them were met with both active and passive resistance.[22]

Tuberculosis control also precipitated such history from below. Early municipal efforts to sanitise the shophouses were frustrated by what Brenda Yeoh terms 'Sisyphean tactics' – evasion and passive resistance.[23] Ultimately, public housing, supported by a system of master planning and enabled by a series of devastating kampong fires after the war, emerged as the dominant housing form in Singapore and won over most people's living preferences.[24] But as the greater part of tuberculosis control took place outside the government hospital, there remained an unremovable degree of social evasion and resistance. Even as the state's efforts reached schools, rural areas and people's homes, they encountered a host of responses from patients and their close contacts (generally household members or those who work closely with the patients in the workplace): refusal or inability to undergo X-ray screening, attend outpatient clinics or complete the lengthy course of antibiotics.

More recently, the use of directly observed therapy in the STEP programme revealed how difficult it was for the government to monitor and change patient behaviour even in a small and regulated city-state like Singapore. Conversely, despite decades of intensive public health education, particularly successful in the campaign against public spitting, exaggerated fears of infection forged a persistent social stigma against sufferers of tuberculosis. The rehabilitation of chronic and cured patients has not been as traumatic as that of sufferers of another dreaded disease – leprosy – but the development of drug-resistant tuberculosis may yet worsen it.[25]

The other strand of the social history was SATA, a non-governmental association formed in 1947 in response to the colonial disinterest in combating tuberculosis. Led by doctors, businessmen and community leaders and supported by volunteers and donors, SATA was among the dynamic social movements which emerged in post-war Singapore. It concentrated on anti-tuberculosis work in the community, such as X-ray screening and public education. The association collaborated with the colonial and PAP administrations, supporting their health policies and becoming mostly a medical service provider, especially after 1965. As a 1997 commemorative publication stated, its founders and leaders were both 'men of vision' and 'pragmatists' who decided to focus their efforts on a serious disease of the general population.[26]

While not a classic civil society organisation in the Western sense, SATA showed the possibilities (and limits) of social activism in Singapore.[27] It took independent, influential positions on important substantive issues such as the need for a centralised registry of tuberculosis cases (which was nevertheless implemented by the colonial government) and for compulsory BCG vaccination (which was shelved by the PAP). SATA was not always successful – or correct – in its stances, but debating the disease, as our book suggests, was germane to the history of tuberculosis in Singapore. Tuberculosis control in this sense was a product of the community as it was of the state.

The shape of the book

The policy and social history of tuberculosis unfolds across nine chapters in our book. The first three trace the contours of the early colonial response prior to the Second World War. Chapter 1 examines the weak, ambivalent – but nonetheless significant – relationship between tuberculosis and the community-run pauper hospital, Tan Tock Seng Hospital, in the second half of the nineteenth century following the transfer of the Straits Settlements to the Colonial Office in 1867. Chapter 2 explores the influence of modern sanitary science, leading the colonial government to take a growing interest in tuberculosis among Chinese coolies residing in the town's shophouses in the early twentieth century. Chapter 3 pursues this interest in both its physical and socio-intellectual manifestations in the 1920s and 1930s, as the colonial state began to build the first public housing in Singapore and to appraise the disease in starkly racial terms; both developments had long-term implications for the governance of the city-state.

The next trio of chapters explore the emergence and expansion of a national tuberculosis policy as part of the decolonisation and political development of Singapore after the Japanese Occupation. Chapter 4 traces the social and political struggles which moved the British colonial government to reluctantly adopt a definitive policy against the disease in 1948. Next Chapters 5 and 6 uncover the implementation of this policy across the colonial and postcolonial divide: they reveal how the PAP government crucially drew upon and expanded the groundwork laid by colonial precedents, particularly in the formation of a centralised government body for the disease, called the Tuberculosis Control Unit, in 1957.

The last three chapters delve into the social dimensions and impact of the anti-tuberculosis programme between the 1950s and 2000s. Chapter 7 peers beyond the sanatorium to map the state's complex and wide-ranging work in the general population, while Chapter 8 looks at the instruments of tuberculosis control for infants and young children, which reached and affected parents, homes and schools. The final chapter on SATA discusses the changes and continuities in the role of the national anti-tuberculosis association in the community. The three chapters throw light on the ways tuberculosis control helped confer and define citizenship in Singapore. They also underline the limits of state policy and highlight people's diverse experiences of, and responses to, tuberculosis control.

Notes

1 Speech by Lee Kuan Yew, Secretary-General of the People's Action Party and Prime Minister of Singapore, on 22 September 1963 after Announcements on the General Election Results, p. 5, www.nas.gov.sg/archivesonline/speeches/record-details/7409855 f-115d-11e3-83d5-0050568939ad.
2 Alison Bashford, *Imperial Hygiene: A Critical History of Colonialism, Nationalism and Public Health* (Houndmills, Basingstoke: Palgrave Macmillan, 2004).
3 Oral History Centre, National Archives of Singapore, Interview with Tay Chin Tian, Reel 6, 3 November 1989.
4 Michael D. Barr, *Singapore: A Modern History* (London: I.B. Taurus, 2019).
5 Carl A. Trocki, *Opium and Empire: Chinese Society in Colonial Singapore, 1800–1910* (Ithaca, NY: Cornell University Press, 1990).
6 James Francis Warren, *Rickshaw Coolie: A People's History of Singapore* (Singapore: Singapore University Press, 2003).
7 Singapore Municipality, *Administration Report 1933*, p. 5-D.
8 Margaret Jones, 'Tuberculosis, Housing and the Colonial State: Hong Kong, 1900–1950', *Modern Asian Studies* 37, 2003, pp. 653–682.
9 The Singapore population is officially classified into Chinese, Malays, Indians, and Others (the last being members of other ethnic groups such as Eurasians and recent migrants not belonging to the three main groups).
10 Christian W. McMillen, *Discovering Tuberculosis: A Global History, 1900 to the Present* (New Haven, CT and London: Yale University Press, 2015), p. 23.
11 Teemu Ryymin, 'Civilizing the "Uncivilized": The Fight against Tuberculosis in Northern Norway at the Beginning of the Twentieth Century', *Acta Borealia: A Nordic Journal of Circumpolar Societies* 24 (2), 2007, pp. 143–161.
12 John Welshman, 'Tuberculosis, "Race", and Migration, 1950–70', *Sociology of Health and Illness* 22 (6), November 2000, pp. 858–882.
13 Emily K. Abel, *Tuberculosis and the Politics of Exclusion: A History of Public Health and Migration to Los Angeles* (New Brunswick, NJ: Rutgers University Press, 2007).
14 Bashford, *Imperial Hygiene*.
15 Lee Chien Earn and K. Satku, *Singapore's Health Care System: What 50 Years Have Achieved* (Singapore: World Scientific Publishing, 2015).
16 Sunil S. Amrith, 'In Search of a Magic Bullet for Tuberculosis: South India and Beyond, 1955–1965', *Social History of Medicine* 17 (1), 2004, pp. 113–130.
17 McMillen, *Discovering Tuberculosis*, p. 59.
18 McMillen, *Discovering Tuberculosis*, p. 66.
19 Ji Han Kim and Jae-Joon Yim, 'Achievements in and Challenges of Tuberculosis Control in South Korea', *Emerging Infectious Diseases* 21 (11), November 2015, pp. 1913–1920.
20 McMillen, *Discovering Tuberculosis*.
21 Kah Seng Loh, 'Within the Singapore Story: The Use and Narrative of History in Singapore', *Crossroads: An Interdisciplinary Journal of Southeast Asian Studies* 12 (2), 1998, pp. 1–21.
22 Brenda S.A. Yeoh, *Contesting Space in Colonial Singapore: Power Relations and the Urban Built Environment*, 2nd edition (Singapore: Singapore University Press, 2003); Kah Seng Loh, *Squatters into Citizens: The 1961 Bukit Ho Swee Fire and the Making of Modern Singapore* (NUS Press and Asian Studies Association of Australia, Southeast Asia Series, 2013).

23 Yeoh, *Contesting Space in Colonial Singapore*.
24 Loh, *Squatters into Citizens*.
25 Kah Seng Loh, *Making and Unmaking the Asylum: Leprosy and Modernity in Singapore and Malaysia* (Petaling Jaya: SIRD, 2009).
26 Lim Kay Tong and Mary Lee, *Fighting TB: The SATA Story (1947–1997)* (Singapore: Singapore Anti-Tuberculosis Association, 1997), p. 9.
27 Liew Kai Khiun, 'Myths of Civil Society and its Culture Wars', in Kah Seng Loh, Thum Ping Tjin and Jack Chia (eds.), *Living with Myths in Singapore* (Singapore: Ethos Books, 2017), pp. 203–212.

1   The pauper hospital

The long-documented history of tuberculosis in Singapore began only some 50 years after Stamford Raffles first founded a factory on the island in 1819. This was after the colony of the Straits Settlements, comprising Singapore, Penang and Malacca, was transferred from the British colonial government in India to the Colonial Office in London in 1867. The transfer established a direct line of command between the metropole and the colony, laying the basis for a more active policy towards sickness and health. More broadly, this change was part of the expansion of modern medical administration throughout the British empire in the second half of the century.[1] Nevertheless, the Colonial Office, like its predecessors in Singapore, soon deemed tuberculosis to be of secondary concern.

It is difficult to write the history of tuberculosis in Singapore in the second half of the nineteenth century. It was a serious disease connected, on the one hand, to immigration and an economic system predicated on free trade, and on the other to the endeavours and struggles of the emergent colonial state. This largely limited tuberculosis to being written from the vantage point of the colonial official and hospital. The post-transfer British administration was more concerned about other diseases. The pages of the Health Department reports were filled with deliberations over dreaded notifiable epidemic diseases such as cholera (outbreaks of which frequently struck the town area, including the hospitals), bubonic plague (leading the government to institute quarantine measures for immigrants) and smallpox (which precipitated a not-too-successful vaccination programme in 1869). The official gaze on non-epidemic illnesses fell on beri beri (a leading killer thought then to be contagious), leprosy and mental illness (both of which provoked social horror and caused sufferers to be kept away from the public in colonial asylums), and the venereal diseases (the focus of which targeted prostitutes under the Contagious Diseases Ordinance).[2]

It was not that tuberculosis was unknown to colonial officials and doctors in late nineteenth-century Singapore, as it was categorised under a group of 'constitutional diseases' or 'fevers'. But while there were fairly periodic, and occasionally substantial, statistics and commentary on other diseases, information on tuberculosis was surprisingly fragmentary, irregular and incomplete. No policy as such existed for the disease. Following the documentary trail of this elusive killing affliction was frustrating: for instance, the yearly medical reports of Tan Tock Seng Hospital (TTSH), the pauper hospital of nineteenth-century Singapore, contributed in large part to the narrative recounted here, but the history of the hospital does not adequately cover that of the illness. It was also difficult to map trends over time as the chroniclers of health and disease in colonial Singapore changed their minds over the diseases to document and highlight – seemingly arbitrarily.

All these reflect the limited interest of Singapore's officials and physicians in tuberculosis up to the end of the nineteenth century. This was indicative of a constrained approach to medical governance, which focused on immediate or manifest threats such as the notifiable epidemic diseases, or illnesses that provoked social anger. Nevertheless, despite the paucity of sources, it was still possible to pin down tuberculosis to some extent and write about it as a historical subject. There was some early official interest in the illness, which grew and was eventually transformed by the emergence of sanitary science at the end of the nineteenth century. The keys to understanding tuberculosis in this early period were the paupers who comprised the bulk of the sufferers of the disease, and the hospital where they arrived (and often died), or which they tried to avoid – TTSH. The paupers and pauper hospital provide a rare glimpse into the underside of an otherwise thriving, globally connected colonial entrepôt which was open to trade, immigrants and sickness. The historical insights are still relevant to contemporary Singapore.

The pauper problem

Until after the transfer of the Straits Settlements, most tuberculosis patients in the colonial record were not paupers, but patients of the General Hospital (GH) who were admitted on account of their illness. GH is usually accepted as Singapore's first modern hospital, built in 1821 and sited in the nineteenth century at various parts of the town area such as Bras Basah Road, Pearl's Hill, Kandang Kerbau, and Sepoy Lines. However, it was originally intended as a hospital for European seamen and employees of the colonial service, both European and Asian. It was from these two groups that GH's tuberculosis patients were drawn, the number of which was very small up to 1870, usually less than ten cases yearly.[3] In the 1850s, local English-language newspapers carried advertisements for 'coughing lozenges' and herbal preparations as remedies for pulmonary tuberculosis, targeted at European and educated Asian consumers.[4] Besides GH, tuberculosis also occurred among the patients and inmates of other institutions in Singapore such as the mental asylum. In 1864, it was also cited among the prevalent diseases at the hospital for convict workers who had been brought from India to carry out public works.[5]

The disease treated at GH was largely transnational, and deadly. In 1873, six out of eight Europeans admitted for pulmonary tuberculosis at the hospital died; among them were Spanish sailors en route to Manila. Their length of treatment lasted 'barely four days', 'having been sent to Hospital from different ships almost in a dying state', while the 'exertion consequent upon their removal to

Hospital hastened their death'.[6] The number of Asian tuberculosis patients at the hospital was small (only two), but the mortality rate was 100 per cent (both patients died). At GH that year, 'The greatest mortality occurs under the head of Phthisis', but the six deaths there were too small for the total of Asian tuberculosis sufferers in Singapore.[7]

In contrast, the bulk of Singapore's population who had tuberculosis represented a much larger and very different group and social class: namely, the paupers. Historically, paupers are interesting: in nineteenth-century England, for instance, pauperism came to the state's notice as a serious problem of work-shy and idle people which was infecting and demoralising the labouring class as a whole. It began to attack the broad charity which hitherto had been given freely to paupers, by distinguishing between those deserving of relief and others who were not. Part of the impulse was punitive: able-bodied paupers were sent to the workhouse under the provisions of the Poor Law as punishment for their idleness, and strenuous efforts were made to screen out 'clever paupers' who were trying to obtain unmerited charity.[8] As a former vagrant in Britain reflected, the twentieth-century establishment still regarded vagrants and paupers as a group of the 'socially sick', whom it found embarrassing and wanted to integrate into capitalist society.[9] The imposed solutions to pauperism were thus thrift and regular hard work. On the other hand, deserving paupers would receive humanitarian relief. They included infirm, decrepit or ill persons who were physically unable to work and provide for themselves, who would be sent to hospitals to be cared for.

Singapore had paupers since its founding, and most of them were sickly. In fact, they were closely tied to its early history. The British government, educated Europeans and Asian community leaders were aware of them and perceived them to be a social issue. As the port of Singapore grew, so ill paupers begging and in many cases dying in the streets became a common sight. Given the lack of effective control over the coastlines of what was an open free port, Singapore's beaches were a convenient dumping ground for the unwanted 'social rejects' of the neighbouring Dutch East Indies and Malay states throughout the nineteenth century. A letter to a local newspaper in 1830 lamented:

> As if we had not a sufficient supply of such living corruption continually imported from various parts of China in junks, we are complemented with all those who find their way to Sambas and Pontiana, who are landed at Palembang and Banca, of which place, Rhio [all places in the Dutch East Indies], is the grand depot. All of them are collected by our kind neighbour and during the night put on shore to the eastward of this island in droves of 20 and 30, who in course of time drag their filthy carcasses to Town, bringing in their train dire diseases with all the commitments of a leper house.[10]

As the letter intimated, many of the paupers had been discarded in Singapore because they were decrepit or ill. Some suffered from leprosy and mental illness – diseases with visible symptoms that provoked public horror or pity. The British discovered most of them to be recently arrived immigrants who spoke in unintelligible tongues, likely Chinese. As the paupers appeared in the streets and other public places – without abode, penniless, seeking alms, and often seriously sick – the authorities deemed them to be illegal vagrants and had them summarily rounded up by police.

The question of what to do with these seized paupers preoccupied the government and educated public. In contemporary British thinking, most of them were deserving of relief. Their diseases and general poor constitution meant that a hospital of sorts would be the answer, but this would not be a hospital in the modern sense of the word, as the 1830 letter made clear:

> there can surely be but one opinion, i.e. that the Chinese and they only be compelled to provide a proper asylum and support for their suffering countrymen.[11]

Thus, the institution for the abandoned paupers was not only a hospital in terms of the therapy it provided, but also a sanctuary, lock-up and death house combined. Such a hospital was meant to keep the paupers from returning to the street – and thus from the public – as much as it was to offer them relief. Just as importantly, as the letter proposed, the hospital would be funded by the Chinese community, rather than the colonial administration. Efforts to deal with the paupers throughout the nineteenth century stem from these two perspectives.

The reference to leprosy in the letter is telling, for the paupers were in equal measure pitied and feared for their diseases. As late as the 1890s, many of the admissions to the Pulau Jerejak leprosarium to the north of Singapore were Chinese who did not speak a word of Malay, having been dropped there from Java, Sumatra or directly from China.[12] Another disease similar to leprosy, also invoking public disgust and social stigma, was mental illness. The sufferers of both diseases were largely paupers. In 1893, the medical superintendent of the mental asylum protested about the island being used as a 'dumping ground of the maimed, the halt, and the blind of our neighbours'.[13] The treatment of these mental sufferers was also deemed less important than their removal from the public eye by police, thereafter which they were held in the convict gaol.[14]

In addition to leprosy and mental illness, tuberculosis was a major disease among the paupers of Singapore in the late nineteenth century. From 1874, paupers suffering (and dying) from tuberculosis began to appear in the official records of Tan Tock Seng Hospital, or the 'Pauper Hospital' as the British aptly called it. TTSH had been established in 1844 at Pearl's Hill to care for 'the diseased of all countries'.[15] The hospital is well-known and highly regarded in Singapore history, although there is as yet no academic study on it. Existing accounts mostly fall in the realm of public history, highlighting the charitable early pioneers and blending the histories of the hospital and the nation.[16] Many aspects of this narrative are valid: TTSH was named after a prominent Straits-born Chinese businessman who had donated the full sum of $5,000 to build it. It was also supported (though at times barely) by subscriptions from Asian businessmen and community leaders, which were supplemented by limited funding from the British government. TTSH was a humanitarian response to the problems of poverty and disease in early colonial Singapore. This was remarkable given that most of the paupers were unwanted, often dying, immigrants. Early in 1843, in envisioning the purpose of the pauper hospital, Tan Tock Seng had written to Governor W.J. Butterworth about the need for

> some place of refuge for the number of my unfortunate countrymen who, at present, while suffering under loathsome diseases crowd the streets of the Town and daily obtrude themselves on the public charity having no other means of obtaining relief.[17]

Thus the paupers with their debilitating and 'loathsome diseases' reveal how ordinary lives and fates were bound up with the historic role of colonial Singapore as a free port and open city with no real borders to the entry of immigrants (and their diseases). Paupers afflicted with leprosy and mental illness had been rendered outcasts because their diseases elicited social stigma and abandonment by family and community. Tuberculosis was rather different, but its sufferers had similarly been deemed extraneous to the colonial society of Western capitalism who now had to be cared for. From the 1870s, tuberculosis sufferers were also dumped – infirm and unwanted – onto Singapore's shores from neighbouring islands in the Dutch East Indies and Malay peninsula, and they were likewise arrested for public vagrancy and sent to a pauper hospital to be treated or, as was more often the case, to die.

The pauper problem opens a path to the history of tuberculosis. Throughout the nineteenth century, British response to tuberculosis was shaped by their thinking on the pauper problem. This allows us to track the illness through the pauper hospital and the colonial archive, but it also raises a historiographical issue: that we may need to look upon the hospital in a different light. In an important sense, the issue of pauperism complicates the history of TTSH and extends it beyond the pioneer narrative. Narrating tuberculosis through the hospital takes into account what being a pauper meant in the historical context, and in terms of how they were perceived and treated as a social problem.

Notwithstanding how sickly and infirm paupers were deemed to merit official assistance, the social history tells a rather different story. The intended beneficiaries of the pauper hospital, who were mandated by law to be held in what was imagined to be a 'refuge' and 'sanctuary', had no say in the matter. But TTSH's patients often acted autonomously with their own feet and left. In 1862, the government reported

> considerable difficulty having been experienced in preventing the patients [of TTSH], who suffer principally from sores and ulcers, from infringing the Hospital regulations and proceeding into the Town for the purpose of begging.[18]

The government instructed the police to round up these alms-seeking absconders from the hospital before the magistrate. Some of them were sent to the House of

Correction in the Convict Hospital for vagrancy, where they were placed on the sick list, and sometimes died soon after.[19]

Was there, then, a real difference between able-bodied and decrepit paupers, or between the workhouse for the former and the hospital for the latter? The ambivalent role of the pauper hospital complicates the work of the social historian. Because of absconding patients and their attempts to avoid arrest, the history of tuberculosis supersedes the hospital, and vice versa. TTSH was concerned with myriad deadly diseases, and we know little of its tuberculosis pauper-patients in the first two decades of its founding, or even if there were any such patients. It was only with progressive state intervention into matters of life and death in the Straits Settlements after 1867 that we gain a clearer picture of a rudimentary colonial policy towards tuberculosis. Yet, the very nature of the pauper policy, based on removing social outcasts from the streets and maintaining them in a hospital until they recovered (or died), means that a large part of the history of tuberculosis exists outside of the institution.

There is a further complication in Singapore's context: the pauper response here was shaped by precedents in Britain, but the practice and implementation were never as fully realised. Prior to the transfer of the Straits Settlements to London, the government of Singapore was dominated by the commercial interests of the East India Company. What little efforts that were made in the realm of social welfare and public health were chiefly meant for the well-being of Europeans and the administrative and military personnel. These endeavours were funded not by custom duties on trade (there were none), but through tax farms which taxed items widely consumed by the Asian general population, such as pork, opium and toddy.[20]

The East India Company was therefore half-hearted about supporting TTSH's operating costs. Its indifference reflected a governing imperative for a port city whose main role was to facilitate the lucrative movements of raw materials and manufactured products between Asia and the West. With economic interests in mind, the British government repeatedly made the argument that more funding for the hospital would discourage subscriptions from the Asian communities, and also increase the dumping of paupers in Singapore. It was only with much reluctance that the government had agreed to supplement the Asian subscriptions. TTSH was badly underfunded and overcrowded throughout its early history, and the living conditions for patients were insanitary, grim and dangerous.[21] In 1861 the hospital was rebuilt at a larger premises at the intersection of Serangoon Road and Balestier Road to accommodate more patients. But for another decade, the only tuberculosis patients in the official record were at GH and these were not paupers.

Tuberculosis at the pauper hospital

The modern history of tuberculosis among the paupers only really began after the transfer of the Straits Settlements to the Colonial Office in 1867. This prompted greater British interest in matters of social welfare and public health in the colony, although partly as a way to fortify colonial authority. There was, in addition, a new concern about the health of the general population, beyond that of the European and Asian urban elites. To fund its increased spending, the colonial government revived the pork tax and introduced a new tax on gambling.[22]

Specifically, the British began to develop a consolidated policy for the paupers and the pauper hospitals. The result, as we shall see, was that TTSH became the institutional focal point for the treatment of paupers afflicted with tuberculosis and other 'worrisome diseases'. The expansion of governance at the imperial level connected with a sustained chorus of local criticism of the government's failure to better support the pauper hospitals of the colony. The metropolitan and local influences belatedly moved the British government to adopt a concerted policy towards sickly paupers, who were divided into three groups: those with leprosy, mental illness and 'other diseases'. Tuberculosis, at this stage in the late 1860s, was not yet named on an equal standing with the first two diseases.

In 1869, the Secretary of State for the Colonies Earl Granville pressed Governor Harry Ord on the 'adoption of some plan' for the 'organisation and future regulation of the Pauper Hospitals' of the Straits Settlements.[23] Ord proposed to the Asian unofficial members of the colony's Legislative Councils that the government take over the pauper hospitals and appoint 'a body of Visitors' to run and manage them. But while those in Penang and Malacca were more agreeable, the Singapore unofficial members opposed the plan as it would incur greater costs in running the hospital.[24] Two years later, in 1871, Ord appointed a four-member Select Committee on Pauper Hospitals, led by Arthur N. Birch, to study the pauper problem in the Straits Settlements. The committee was instructed to consider the amount of accommodation in the colony's pauper hospitals required to house patients suffering from leprosy, mental illness and 'other diseases'. It was also tasked to inquire into the question of the government taking over these institutions. Searching for a consolidated policy, the committee would 'consider how far it is practicable to treat the Colony as a whole, and to bring together, from the three Settlements, certain classes of cases for treatment in one place'.[25] However, its eventual report to the Legislative Councils, submitted in 1871, proposed different approaches for the three categories of disease.

For pauper-sufferers of mental illness in the Straits Settlements, the committee advised that they should be kept, as before, in an asylum located in Singapore, with only slightly increased accommodation for ten additional patients (totalling 150). But the committee had a different view of leprosy. Although the colonial government had in the same year opened a new asylum (completed earlier in 1868) for 140 patients at Pulau Jerejak, the report proposed that each Settlement should have its own 'island asylum' like Pulau Jerejak, located close to the town area. The committee also made more expansive recommendations for accommodation for leprosy: unlike mental illness, it urged the government to build accommodation for 100 leprosy patients in Singapore (a substantial increase from the existing 20), 60 in Penang and 20 in Malacca. At the time, Singapore's leprosy patients were housed in an overcrowded ward inside TTSH. Subsequent to the report, the government periodically moved some patients to Pulau Jerejak to relieve overcrowding, but most would remain in Singapore, albeit on the main island and not on a 'leprosy island' as suggested by the report.[26]

For paupers with 'other diseases', the committee envisaged their continued treatment in the various pauper hospitals of the Straits Settlements. It proposed additional accommodation for a total of 600 patients in Singapore, 300 in Penang and 100 in Malacca. This meant, for Singapore, a substantial increase of accommodation for 420 patients from the existing 180. The increase would entail the expansion of TTSH and also changed the size and composition of its patient population. The report proposed the expansion of TTSH at its current site at the intersection of Serangoon Road and Balestier Road, rather than the building of a new hospital. The building of additional accommodation progressed slowly: three temporary sheds with 180 beds were completed at TTSH only three years later in 1874, raising the accommodation to 356. In 1879, a further three wards were added, bringing the accommodation to 552 – still below the suggested number. TTSH remained overcrowded.

From the 1870s, the hospital was reaching the point of severe overcrowding. The bulk of these patients were working-class immigrants from south-eastern China. In 1873, TTSH admitted 1,461 patients and although some of them died soon after arrival, the number was already more than double what the committee had projected. The number of admissions jumped to 2,757 in 1877, hit 4,000 in 1882 and exceeded 7,000 in 1897. The substantial increases were mostly due to the immigration-led expansion of Singapore's population in the final decades of the nineteenth century. The island's population rose by 136 per cent from 96,000 in 1871 to 227,000 in 1901, but even this was less than the four-fold surge in TTSH's patient population over the same period. Most of the increase in Singapore's population was due to immigration: from 1881 to 1901, the natural change was a decrease of 73,000, but this was more than compensated by a net migrational increase of 163,000.[27] Similarly, a significant proportion of TTSH's patients were from the region; in 1884, a quarter of the patients were migrants from plantations in Johor to the north of Singapore.

For the select committee, TTSH would not only expand physically but also administratively, although this turned out to be a much more fractious issue. On the question of state funding and management, the committee dismissed the commonplace objection that both the Dutch East Indies government and locals would be encouraged to dump sickly immigrants in Singapore. Yet, the committee advised the government against taking over the pauper hospitals of the Straits Settlements. The current system should continue, it argued, with the Asian communities collecting and administering public subscriptions and the government providing supplementary grants and advice by its medical officers. The committee repeated the old argument that Asian subscriptions would fall off if the government took over the hospitals.[28]

The Secretary of State for the Colonies, Lord Kimberly, disagreed with the committee, deeming that it had not gone far enough in providing for the care of the paupers. Although he left open the question of whether the government should care for all paupers or only the sickly ones, he insisted that 'the local Government should now take upon itself effectively the discharge of this duty', and that 'H.M.'s Government must hold the Governor responsible for giving effect to its instructions'. The hospitals, he added, should be supervised by an officer of the government who would regularly inspect them and who would be directly responsible to the Governor.[29] In 1873, concurrently with the physical expansion, TTSH was handed over to a Committee of Management that was in part composed of British officials: namely, the Colonial Secretary, Principal Civil Medical Officer, Inspector-General of Police, and Protector of Chinese. But it also included prominent Asian, mostly Chinese, businessmen and community leaders, such as Tan Kim Ching (the eldest son of Tan Tock Seng), Tan Seng Poh, Tan Beng Swee, Cheang Hong Lim, and Hoo Ah Kay (also known as Whampoa). In 1880, the Tan Tock Seng Hospital Corporation was formed as a combined venture of the British government and Asian communities to continue to care for sickly paupers.

With the expansion of TTSH in the 1870s, we may chart the history of tuberculosis in Singapore a little more clearly. One apparent finding is the marked increase in mortality rates for all diseases and for tuberculosis in particular, as greater numbers of paupers who were in advanced or terminal stages of their illness were admitted to the hospital. Prior to the transfer of the Straits Settlements during 1861 to 1867, the monthly mortality rate for all diseases at TTSH ranged from 3.7 per cent to 8.4 per cent of the patient population.[30] However, in 1873 the mortality rate surged to over one-fifth of the admissions, and to one-third in the following year with the completion of the three additional wards, before coming down to one-quarter in 1876. Thereafter till 1894, the mortality rate hovered between 11 per cent and 18 per cent. In the 1880s, the authorities pointed to lower daily numbers of patients treated and periods of stay at TTSH as evidence that locals were seeking treatment at earlier stages of their illness.[31] Such a claim was premature, however, as death rates exceeded 20 per cent in the late 1890s, reaching a high of 29 per cent in 1896.

The high mortality rates at the nineteenth-century hospital reinforced its unpopularity with locals with their own institutions and cultures of healing. To them, the supposedly modern hospital was not a haven or place of healing, but a final destination for the doomed. This was true of other hospitals as it was of TTSH. The British observed of even the better-off classes of patients of GH in 1872, 'This Hospital is not generally sought after by residents of the place who have a strong aversion to coming in as patients. The few Government clerks who were admitted were compelled to come.'[32]

Over time, GH played a lesser role in the treatment of tuberculosis. In 1874, it admitted seven European and five Asian patients with pulmonary tuberculosis, compared to 50 Asians suffering from the same disease at TTSH. In subsequent years, European tuberculosis patients continued to be treated at GH; in 1883 three Europeans died there from the disease. The vast majority of Asians with tuberculosis were treated at TTSH, although up to the early 1900s, there were still small numbers of Asians admitted to GH with the disease. But generally, the Asians admitted to GH consisted of a narrower group of patients employed by the colonial establishment or sent by their employers, namely, 'servants, dockyard workers, native seamen, and the cases taken up by the Police, most of whom are the subjects of accidents', and policemen (there was a police ward in the hospital).[33] In 1895, some of these admissions were self-paying customers who were fairly well-off.[34] They were also often going for treatment for less serious illnesses.[35] In 1890, 14 Asian patients suffering from tuberculosis were admitted to GH, of whom four died. In 1905 and 1906, the number of Asian admissions for tuberculosis at GH increased to 77 (including 44 Chinese), compared to 21 in 1891–1892. Among them were 11 rickshaw pullers and nine new arrivals from China.[36]

In contrast, there were 379 admissions for tuberculosis to TTSH in 1905.[37] Since its expansion in 1874, the hospital had become the chief institution for treating Asians with the disease. Tuberculosis (or more accurately, 'phthisis pulmonalis', 'pulmonary consumption' and 'tubercule of lungs') quickly became one of the recognised killing diseases there. In the 1880s, it began to be included in the medical reports within the group of major killers such as beri beri, malaria and dysentery. In that decade, the number of deaths from tuberculosis at the hospital rose from 42 in 1878 into the 70s and 80s. In 1890, tuberculosis killed 100 persons at TTSH, compared to 161 who died from beri beri, the top killer. In 1899, the number of deaths from tuberculosis jumped to 190 and again to 276 the following year.

In addition to the number of deaths, the mortality rate at TTSH for Asians with tuberculosis, a large number of whom were paupers, was probably high, although its certainty is hindered by the lack of regular statistics for tuberculosis. In 1874, there were totals of 50 admissions to TTSH and 43 deaths, in 1877, 75 admissions and 54 deaths, and in the following year, 55 admissions and 42 deaths. We are then without comparable statistics until 1884, when 134 admissions and 81 deaths were recorded, followed by a further lack of official data. In 1895, however, TTSH documented the treatment of 142 tuberculosis patients and 116 deaths. The statistics in Singapore paralleled those of the Straits Settlements which appeared in the late 1890s, where the number of admissions and deaths of tuberculosis patients also increased, respectively, from 519 and 296 in 1894 to 784 and 440 in 1900 – rises of 51 per cent and 49 per cent respectively.

The composite picture of the tuberculosis patient at TTSH was a working-class Chinese man, although not all of them were paupers. The presence of the Protector of Chinese and a large number of Chinese representatives in the Committee of Management suggests that a disproportionately large majority of patients at TTSH were Chinese. Most of them hailed from the waves of Chinese migrants of peasant background who arrived in Singapore and Southeast Asia in large numbers from the 1870s. In the decades between 1871 and 1901, the Chinese composition of Singapore's population rose from 58 per cent to 72 per cent.[38] Yet the proportion at TTSH was much higher. In 1882, 95 per cent of all patients at TTSH were Chinese. This suggests that non-Chinese sufferers – Malay, Indian and other ethnic minorities – more rarely sought treatment at the hospital, other than the employees of the colonial civil service. The Malays in particular, being a more settled group, would have received family and community support within their own settlement.

In addition, the mass immigration of Chinese to Singapore suggests that an increasing number of tuberculosis patients at TTSH were not paupers in the traditional sense of the term. These Chinese would have been of working-class background – coolies, *sinkeh* (new arrivals) to more established Chinese – but they were not destitutes without the means to help themselves or local kin to assist them in finding lodging and employment. On the contrary, they usually arrived on a chain of migration, following a brother or kinsman from their home village to Singapore. These coolies would likely have entered the hospital only if they had exhausted all other means of treatment.

Although there are no statistics for tuberculosis, sufferers of the disease at TTSH from the 1870s were overwhelmingly male. This followed larger gender trends at the hospital and in Singapore as a whole. Between 1871 and 1901, the sex ratio of Singapore's population was heavily skewed towards men, at around 3,000 males per 1,000 females. However, among the Chinese, the sex ratio was nearly halved from a high of 6,100 to 3,900 males over the same period.[39] In contrast, the sex ratio at TTSH was much more uneven, being linked to the racial composition. In 1890, among the 5,891 patients treated at TTSH, only 59 were women. In 1893, it was a major killer among women who, like the men, were commonly admitted at a late stage of their illness or in poor constitution.[40] Tuberculosis also appeared among the patients of a female ward at TTSH, constructed in 1889. The deteriorating state of the soon overcrowded ward further worsened the women's health in the late 1890s.[41]

The disease in the community

Reading between the lines of the colonial records and statistical data suggests, however, that much of the history of tuberculosis lay beyond TTSH. The hospital's statistics are arguably a barometer of the state of health and illness in Singapore, since the patient population was drawn from the island's working-class population and suffered from the same diseases that were the major killers at the national level.[42] Yet the statistics were by no means a perfect measure. For one, they did not cover sufferers who were repatriated from Singapore.[43] There was also a major inherent uncertainty in official figures, as the cause of death was often not determined by a certified medical officer.[44] Tuberculosis presented a greater problem in this regard than other diseases because, other than among street paupers, typical symptoms such as cough and weight loss over a prolonged period of time were not particularly visible or dramatic. In addition, tuberculosis was often erroneously diagnosed as pneumonia.[45]

There was also a lack of statistics on tuberculosis for the whole of Singapore. The colonial and medical chroniclers of disease in the late nineteenth century did not deem tuberculosis as a disease that needed to be monitored over time outside of TTSH. The 1883 report of births and deaths in the Straits Settlements is both unique and problematic in this regard. It attributes a large figure of 614 deaths to 'consumption' in the colony, compared to a mere 78 deaths at TTSH recorded in the corresponding report on the civil hospitals. In the former report, tuberculosis was also cited as causing one-fifth of all deaths attributed to the major killing diseases of Singapore.[46] The data was for one year only, but it suggests that despite the expansion of TTSH, the vast majority of tuberculosis sufferers did not die – or were even treated – in a hospital. It is possible that these deaths occurred in homes or in the streets.

In subsequent years up until the early 1890s, we are again frustrated by the annual reports failing to provide statistics for tuberculosis for the whole of Singapore. Only the statistics for the illness at TTSH and other pauper hospitals in the Straits Settlements continue to be recorded. There were also figures for the much smaller numbers of inmates of Singapore's prison hospital and mental asylum who contracted tuberculosis, and who were probably also patients of TTSH. The lack of national data seems inconsistent with the trends observed at the hospital. There is also another interesting point about the data: at TTSH, tuberculosis was consistently mentioned as one of the major killing diseases, but not as a disease with the highest patient admissions. This points, as we noted, to a high mortality rate, with patients admitted at advanced or terminal stages of the disease. The fact that tuberculosis sufferers avoided the hospital (or arrest by police) until it was too late again attests that the institution was unpopular with Asians.

Absconsion from TTSH throws further light on the ambivalent relationship between hospital and patient. There is no specific data on the absconsion of tuberculosis patients from the hospital, but we may surmise that the phenomenon cut across categories of disease. Hospital statistics reveal that TTSH could not prevent patients from escaping for most of the late nineteenth century even as it undertook to care for increasing numbers of paupers. In 1874, when the hospital admitted a total of 1,180 patients, 399 of them died, while an even greater number – 467 (or 40 per cent) – absconded. In 1876, the number of escapes rose to 574 (46 per cent), but plummeted curiously in the next four years (with only an impossible nine escapes in 1878). From 1882, the numbers rose again. The following year, when 309 patients escaped, the authorities blamed the size of the hospital grounds and the lack of police security (which was placed only at the front gate). Allegedly, the problem was particularly serious at night, and the formation of night patrols to deter the patients was called for but never implemented.[47] In 1884, the hospital took a different approach, dismissing the police guard and adding iron bars to windows and fastenings to doors, but the number of escapes leaped to 470, one-tenth of the patient population.[48]

Thereafter the number and proportion of escapes fell, though possibly not the desire to leave. In 1886, a new iron fence was installed which sliced the number of absconsions from 261 to 56. In response to still some 'patients burrowing at night under the fence like rabbits', the hospital placed concrete structures over the soil, which was effective as a further deterrent.[49] But the number of absconsions rose again in subsequent years to over 100 until 1895, reaching highs of

224 in 1888 and 254 and 1889. From 1895, the number fell to a low of 43 in 1896 but jumped up again to 283 in 1900. These figures do not constitute more than 5 per cent of the total patient population (and was sometimes much lower), but they plausibly reflected the hospital's ability to physically frustrate escape more than the desire of patients to remain.

How did the hospital account for these possibly deflating incidents? The 1883 medical report blamed escapes on Chinese custom – the practice of alms-giving which lured destitutes from sanctioned safe havens such as hospitals and asylums.[50] This hardly placed the role of TTSH as sanctuary and refuge in a good light. But the British also sought to highlight and advocate the appeal of their institutions: they made the distinction between working-class coolies who voluntarily sought treatment at the hospital and rarely absconded, and the ostensibly smaller group of 'professional beggars' brought in by the police, who often did. The administration noted that the beggars' motive was plainly financial, as they usually left the day after admission with their hospital clothes and blankets (and sometimes those of other patients).[51] The reports also claimed that some of the escapees returned after a period of begging or after having made 'small purchases' outside – these, then, were not true absconders but temporary departures.[52]

There is, however, a link between the absconsions and the paupers. We do not know who constituted the professional beggars – did they also include social rejects from the neighbouring countries? In 1880, a letter to the press complained about the many 'Lazars' – sore-covered leprosy sufferers – begging in the streets, and wondered why they were not treated at TTSH.[53] That year, a legislative councillor likewise commented that there were sickly destitutes begging at Collyer Quay and the Esplanade while beds in the hospital lay empty.[54] Were these beggars paupers newly arrived in Singapore or escapees from the hospital? The official records provide no answer.

It seems that the absconsions were also conflated with avoidance of modern medicine. This points to a much larger group of the general population who did not wish to be treated in hospitals. The official notion of absconsion implied that an individual wilfully broke a reasonable agreement to be properly treated at an adequate institution of healing. However, this premise is untenable given that TTSH was at times not a sanctuary, but a congested, dangerous place threatened by diseases such as cholera and beri beri. In 1884, a cholera outbreak at the hospital, which affected 20 patients and killed 16, was traced to insanitary wells there. To uneducated Asians, the hospital also gained ill repute as a place to die, rather than to recover. In this context, the self-willed departure of some patients was a rational choice to leave unsafe places, although their act of agency to resume begging would lead them to be seized again by police and returned to the hospital. This only reinforced the vicious cycle of overcrowding and high mortality at TTSH.

At century's end

Nevertheless, the tuberculosis situation began to change at the turn of the century. The colonial government was being alerted to the seriousness of tuberculosis and its social dimensions beyond the old pauper policy. The early 1890s heralded a change of sorts: in addition to the statistics on the disease at TTSH, the medical doctors were finally commenting on the wider menace it posed. The threat, interestingly, was not first observed at TTSH but in the prison hospital and mental asylum. In 1892, tuberculosis was reportedly on the rise in the prison hospital, with the number of cases doubling to 31.[55] The following year, medical doctors again highlighted it as a major killer at the prison hospital and mental asylum, due to 'some defective sanitary conditions, most probably overcrowding and defective ventilation'. One doctor warned that as most tuberculosis patients in the prison hospital were serving short sentences, the disease was on the rise among the general population of Singapore.[56]

In 1894, the authorities repeated the belief that from the increasing numbers at TTSH, the prison hospital and mental asylum in recent years, 'the disease is more frequently met with now than it was formerly'.[57] In the same year, the Straits Settlements health statistics duly recorded tuberculosis as one of the major illnesses under General Diseases, with 519 admissions and 216 deaths, while excluding it from the class of respiratory diseases.[58] It is not known how many of these numbers were from Singapore, but it is significant that only 102 deaths from tuberculosis were recorded in TTSH. In 1895, the administration declared that the illness was still increasing throughout the Straits Settlements; in that year, the number of tuberculosis deaths at TTSH rose to 116. Medical doctors blamed tuberculosis, arising from poor sanitation, for increased death rates in the prison hospital, with deaths jumping from three in 1889 to 28 that year. They emphasised rural–urban differences in the incidence of tuberculosis, which was especially marked in Singapore.[59] The following year, when the number of deaths from tuberculosis in the prison hospital fell by nearly half to 16, the authorities were puzzled and struggled to explain it.

The references to sanitation are historically significant, as the following chapter will demonstrate. In the official and medical view of tuberculosis, pauperism was giving way to sanitation and environmental science. Earlier, the 1886 medical report had expressed its fear about the connection between sanitation and housing in Singapore, noting how many houses in the urban area were poorly built with few sanitary safeguards, particularly the outhouses. Although the report did not specify tuberculosis – stressing instead water-borne diseases like typhoid – it warned about the deteriorating state of sanitation as the town population expanded and became denser. 'Sanitation', the report emphasised through a religious metaphor, 'is a science, not abstruse, but simple, being nothing more nor less than the science of cleanliness which, according to a well-known adage, ranks next akin to godliness'.[60]

Similarly, in 1897, the Municipal Commission worried about the sanitary factors which had made tuberculosis a leading killer among both young adults aged from 25 to 45 (both males and females) and infants and children from 0 to five. The disease was also responsible for one-sixth of deaths among coolies which, the government warned, highlighted the 'condition of overcrowding, dirt and disregard of Sanitary conditions which predisposes them to attacks of disease and reduces their powers of resistance when attacked'.[61] The previous year, the Commission had appointed two medical men as assistant registrars, who could more accurately establish the incidence of the leading diseases in the town area, including tuberculosis.[62]

Earlier in the nineteenth century, despite the transfer to the Colonial Office, the British colonial government had focused on tuberculosis at the hospital, specifically Tan Tock Seng Hospital; they were content to record the number of patients who arrived or died there, voluntarily or otherwise. By the end of the century, however, the logic of sanitation, triggering concerns about the effects of overcrowding in urban housing, would provide the impetus for the state to pursue the elusive disease beyond the pauper hospital into the homes of the Chinese working class of Singapore.

Notes

1 Lenore Manderson, *Sickness and the State: Health and Illness in Colonial Malaya, 1870–1940* (New York: Cambridge University Press, 1996); Brenda S.A. Yeoh, *Contesting Space in Colonial Singapore: Power Relations and the Urban Built Environment* (Singapore: Singapore University Press, 2003), 2nd edition; David Arnold (ed.), *Imperial Medicine and Indigenous Societies* (Manchester; New York: Manchester University Press, 1988).
2 CO 275/8 Letter from the Commissioner of Police, Straits Settlements, and the Colonial Surgeon on the Subject of Vaccination, 9 October 1868.
3 For a detailed account of the formation and early history of TTSH until the early 1870s, see Lee Yong Kiat, *The Medical History of Early Singapore* (Tokyo: Southeast Asian Medical Information Centre, 1978). The book, however, focuses on administrative and financial matters and does not dwell on tuberculosis.
4 *Singapore Free Press and Mercantile Advertiser*, 9 January 1852.
5 Annual Report of the Administration of the Straits Settlements, for the Year 1864–1865, pp. 67–68, in Robert L. Jarman (ed.), *Annual Reports of the Straits Settlements, 1855–1941* (London: Archive Editions Limited, 1998).
6 CO 275/18 Annual Medical Report of the Civil Hospitals in the Straits Settlements for the Year 1873, pp. xxv–xxvii.
7 CO 275/18 Annual Medical Report of the Civil Hospitals in the Straits Settlements for the Year 1873, p. xxvii.
8 Stedman Gareth Jones, *Outcast London: A Study in the Relationship between Classes in Victorian Society* (Oxford: Clarendon Press, 1971), p. 272.
9 Philip O'Connor, *Britain in the Sixties: Vagrancy* (London: Penguin Books, 1963), p. 18.
10 Quoted in Lee, *The Medical History of Early Singapore*, p. 87.
11 Quoted in Lee, *The Medical History of Early Singapore*, p. 87.
12 CO 275/50 Annual Medical Report of the Civil Hospitals in the Straits Settlements for the Year 1895, p. 608.
13 CO 275/47 Annual Medical Report of the Civil Hospitals in the Straits Settlements for the Year 1893, p. 350.
14 Lee, *The Medical History of Early Singapore*. In 1841, mentally ill people were segregated in the Insane Hospital and then in a larger Lunatic Asylum, established in 1861.
15 Lee, *The Medical History of Early Singapore*, p. 184.
16 See, for instance, Lee Siew Hua, *150 Years of Caring: The Legacy of Tan Tock Seng Hospital* (Singapore: Tan Tock Seng Hospital, 1994).
17 Quoted in Lee, *The Medical History of Early Singapore*, p. 100.

18 Annual Report of the Administration of the Straits Settlements, for the Year 1862–1863, p. 39, in Jarman, *Annual Reports of the Straits Settlements*.

19 Annual Report of the Administration of the Straits Settlements, for the Year 1865–1866, p. 39, in Jarman, *Annual Reports of the Straits Settlements*.

20 Lenore Manderson, *Sickness and the State: Health and Illness in Colonial Malaya*, and 'Health Services and the Legitimation of the Colonial State: British Malaya 1786–1941', *International Journal of Health Services* 17 (1), 1987, pp. 91–112.

21 Lee, *The Medical History of Early Singapore*.

22 Manderson, 'Health Services and the Legitimation of the Colonial State'.

23 CO 275/10 Copy of a Despatch No. 153, dated 4 September 1869, from the Right Hon'ble the Secretary of State for the Colonies, with Reference to the Pauper Asylums and to Pauperism in the Straits Settlements, 15 November 1869, p. cx.

24 CO 275/15 Minute of his Excellency the Governor Covering Correspondence Respecting Lepers and Paupers, 15 August 1872, p. lxxv.

25 CO 275/13 Report of the Select Committee on Pauper Hospitals Laid before the Legislative Council by the Hon'ble the Auditor-General, 5 June 1871, p. lxiv.

26 In 1926, a Leper Asylum was opened at the former Trafalgar rubber estate in Singapore, initially for female sufferers, with a ward for male sufferers added to it four years later. This Singapore leprosarium would indeed begin to receive leprosy patients from Pulau Jerejak and other asylums in Malaya. Kah Seng Loh, *Making and Unmaking the Asylum: Leprosy and Modernity in Singapore and Malaysia* (Petaling Jaya: SIRD, 2009).

27 Saw Swee Hock, *The Population of Singapore* (Singapore: Institute of Southeast Asian Studies, 1999).

28 CO 275/13 Report of the Select Committee on Pauper Hospitals Laid before the Legislative Council by the Hon'ble the Auditor-General, 5 June 1871.

29 CO 275/15 Minute of his Excellency the Governor Covering Correspondence Respecting Lepers and Paupers, 15 August 1872, p. lxxiv.

30 CO 275/8 Return of the Average Monthly Number and Mortality of the Pauper Inmates in Tan Tock Sing's Hospital, from the 1 January 1861 to 30 April 1868, 8 June 1868.

31 CO 275/30 Annual Report of Tan Tock Seng Hospital for the Year 1884.

32 CO 275/17 Annual Medical Report of the Civil Hospitals in the Straits Settlements for the Year 1872, p. xxvii.

33 CO 275/23 Annual Medical Report of the Civil Hospitals in the Straits Settlements for the Year 1878, p. cxvii.

34 CO 275/50 Annual Medical Report of the Civil Hospitals in the Straits Settlements for the Year 1895.

35 CO 275/35 Annual Medical Report of the Civil Hospitals in the Straits Settlements for the Year 1888.

36 W.J. Simpson, *Report of the Sanitary Condition of Singapore* (London: Waterlow & Sons, 1907).

37 CO 275/72 Straits Settlements Medical Report for 1905.

38 Saw, *The Population of Singapore*.

39 Saw, *The Population of Singapore*.

40 CO 275/47 Annual Medical Report of the Civil Hospitals in the Straits Settlements for the Year 1893.

41 CO 275/55 Annual Medical Report of the Civil Hospitals in the Straits Settlements for the Year 1897.

42 CO 275/50 Annual Medical Report of the Civil Hospitals in the Straits Settlements for the Year 1895.

43 D.J. Galloway, 'Notes on Tuberculosis', in Singapore, *Proceedings and Report of the Commission Appointed to Inquire into the Cause of the Present Housing Difficulties in Singapore, and the Steps Which Should be Taken to Remedy Such Difficulties*, Vol. II (Singapore: Government Printing Office, 1918).

44 CO 275/29 Annual Report on the Registration of Births and Deaths in the Straits Settlements for the Year 1883.

45 D.M. McSwan, 'The Problem of Tuberculosis with Special Reference to Singapore', *Journal of the Malayan Branch of the British Medical Association* 1 (3), December 1937, pp. 209–211.

46 CO 275/29 Annual Report on the Registration of Births and Deaths in the Straits Settlements for the Year 1883; CO 275/29 Annual Medical Report of the Civil Hospitals in the Straits Settlements for the Year 1883.

47 CO 275/29 Annual Report on the Registration of Births and Deaths in the Straits Settlements for the Year 1883.

48 CO 275/30 Annual Report of Tan Tock Seng Hospital for the Year 1884.

49 CO 275/32 Annual Report of Tan Tock Seng Hospital for the Year 1886, p. 34.

50 CO 275/29 Annual Medical Report of the Civil Hospitals in the Straits Settlements for the Year 1883.

51 CO 275/29 Annual Medical Report of the Civil Hospitals in the Straits Settlements for the Year 1883.

52 CO 275/28 Annual Medical Report of the Civil Hospitals in the Straits Settlements for the Year 1882.

53 *Straits Times Overland Journal*, 5 July 1880.

54 *Straits Times*, 4 September 1880.

55 CO 275/45 Annual Medical Report of the Civil Hospitals in the Straits Settlements for the Year 1892.

56 CO 275/47 Annual Medical Report of the Civil Hospitals in the Straits Settlements for the Year 1893, p. 350.

57 CO 275/49 Annual Medical Report of the Civil Hospitals in the Straits Settlements for the Year 1894, p. 193.

58 CO 275/49 Annual Medical Report of the Civil Hospitals in the Straits Settlements for the Year 1894.

59 CO 275/50 Annual Medical Report of the Civil Hospitals in the Straits Settlements for the Year 1895, p. 562.

60 CO 275/32 Annual Medical Report of the Civil Hospitals in the Straits Settlements for the Year 1886, p. 26.

61 Singapore Municipality, *Administration Report 1897*, p. 85.

62 Singapore Municipality, *Administration Report 1896*.

2   Disease of town-dwelling Chinese

A nascent colonial policy towards tuberculosis emerged in Singapore at the turn of the twentieth century. This was not so much because of statistics and trends: the incidence and death rates of the disease had already been rising in the final decades of the nineteenth century, making tuberculosis one of the town's major killing diseases. Indeed, concern over tuberculosis grew despite the decline of the mortality rate in the early twentieth century. What changed was that colonial officials and doctors now linked tuberculosis to broader policy questions about housing, the urban labouring class, particularly the Chinese, and the governance of Singapore society. In 1905, the Municipal Health Officer, in a study of autopsies performed at Tan Tock Seng Hospital, surmised that 'the infective bacillus plays sad havoc amongst the Chinese of the coolie class'.[1] Three decades later, the Straits Settlements medical services proclaimed tuberculosis as 'a disease of the town-dwelling Chinese'.[2]

The colonial interest in the paupers of the entrepôt in the late nineteenth century morphed into a concern with environmental sanitation and coolie dwellings. Where the British government had previously been preoccupied with sick destitutes, it now worried about the sanitary states of buildings, cubicles, streets, and back lanes, and working-class denizens. The establishment of the Singapore Municipal Commission in 1887 provided the organisational impetus: the Commission would manage and regulate a wide range of affairs in the town area, not only sickness and disease but also sanitation and housing. The Commissioners comprised both local Europeans and Asians of the elite classes, but the single most important person who initiated the paradigm shift for tuberculosis was W.J.R. Simpson, a foreign medical doctor and sanitary expert. On the invitation of the colonial government, Simpson surveyed Singapore's shophouses in 1906 and wrote an influential report on the need for municipal reform. That it required an outside expert to bring about such a change highlighted the openness of Singapore to metropolitan ideas.

Simpson's report gave the government the conceptual tools to not only control tuberculosis, but also to redevelop the town and manage the urban population. It introduced notions of state-led housing development and urban planning which, in the long run, would transform the physical and socio-economic landscape of Singapore. Just as crucially, the study placed equal emphasis on the health and social habits of the Chinese coolie class. In the early decades of the twentieth century, the administrators and doctors of Singapore concluded that improving the housing would need to go hand in glove with altering the dwellers' mindsets and behaviour.

Tuberculosis and the shophouse

The attention given to tuberculosis may seem rather surprising given one contradictory fact: from a high point in the 1900s, its incidence and death rate declined in the 1920s and 1930s. In 1904, the Straits Settlements administration warned about how tuberculosis, 'having once got a foothold and having found a suitable environment, has steadily spread' among the colony's Asian population.[3] The following year, the government, reporting a big rise in the number of deaths from 2,534 the previous year to 3,015, repeated that 'This disease seemed to be rapidly increasing'.[4] In Singapore specifically, there was a large jump in the number of deaths from 1,644 to 1,946 that year, which was attributed to 'overcrowding and bad sanitation'.[5] As in the previous century, much of the high death rate in the early twentieth century was due to how 'The influx of sick labourers from neighbouring States and islands is considerable and does much to augment our death-rate'.[6] In 1920, the medical services warned of the continuing trend of tuberculosis deaths numbering above 3,000 since three years ago.[7]

Thereafter, however, the severity of tuberculosis appeared to progressively diminish in Singapore and the Straits Settlements. Absolute numbers had risen. The number of tuberculosis admission to hospitals in the Straits Settlements, for which we have the most comprehensive figures, rose from 784 in 1900 to 1,103 in 1910, 1,450 in 1920, 2,230 in 1930, and 3,033 in 1938, an increase of 287 per cent over the period. The number of hospital deaths also increased from 440 in 1900 to 554 in 1910, 698 in 1920, 966 in 1930 and marginally to 1,045 in 1938 (a much smaller rise of 138 per cent). But these figures need to be seen in context due to the increase of the general population over the period, and also to more people seeking treatment at the hospitals. More useful is the death rate for tuberculosis in the Straits Settlements, which counts both hospital and outside cases as a ratio of the general population. In the colony, the total number of deaths from tuberculosis ranged between 2,400 and 3,300 from 1904 to 1938, but the death rate decreased from a high of 4.9 per 1,000 in 1905 to 3.9 in 1910, 2.9 in 1923 and 1.8 in 1932 at the trough of the Great Depression. For the remainder of the 1930s, the death rate was between 1.8 and 2.1.

In Singapore, there was a similar trend despite the Municipal Commission continuing to warn about the effects of heightened overcrowding in the town.[8] The absolute number of tuberculosis deaths was fairly stable below 2,000 throughout the early twentieth century except for one year (2,184 deaths in 1918, likely due to the knock-on effects of the worldwide influenza epidemic on local health).[9] In context, the island's population grew from 226,000 in 1901 to 303,000 in 1911, 418,000 in 1921, and 557,000 in 1931 – more than doubling over the period. The tuberculosis death rate for Singapore dropped from a high of 7.7 in 1905 to 5.8 in 1908, 4.4 in 1912, 3.7 in 1922, and 2.3 in 1924. After this, the death rate stabilised between 2.0 and 2.8 – still higher than the colony's, but much lower than earlier figures. The downward trend was especially marked in the 1920s and 1930s. In 1934, the Municipal Health Officer P.S. Hunter apologised that although he had repeatedly highlighted the tuberculous scourge, he should also have pointed out 'its very pronounced decrease in mortality', by more than half from 1911 to 1931.[10] He did not know the reason, but attributed the outcome to improved water and food supply, sewage disposal, drainage and cleansing which had bolstered people's resistance to sickness.

Are these numbers reliable? There are undoubtedly methodological issues, as was already the case in the nineteenth century. In general, the colonial government had more accurate information on tuberculosis deaths than its incidence in the community. But even the former belie problems in the ways colonial data was generally derived and compiled. As Hunter himself conceded, many officers of the Births and Deaths Department of the government who certified the deaths possessed no medical expertise, but would 'arrive at a diagnosis after hearing the history of the illness from relatives and friends of the deceased and under the circumstances, must perforce take a good deal for granted'.[11]

In 1922, the colonial government reported 3,431 deaths occurring in Singapore's hospitals and 2,909 deaths being certified by medical practitioners, while 5,418 cases were certified by qualified registering officers after the person's death, who had probably been untreated, while 1,898 (or 14 per cent of the total) cases were uncertified.[12] In 1930, 20 per cent of the deaths were uncertified.[13] Although it appears that the bulk of deaths were known to the state, it is plausible that the proportion of the 'uncertified' was much larger, and possibly indeterminate. In 1937, nearly 29 per cent of the deaths were certified merely by 'a view of the corpse'.[14] The other revelation from the figures was the large numbers of the sick who did not bring themselves to a hospital or Western doctor.

How did the colonial doctors and administrators interpret the trends? Occasionally, they acknowledged the fact: in 1932, the authorities admitted that although the statistics might be under-reported, tuberculosis was not on the increase in the colony.[15] They also, as we shall see later, partially endorsed the beneficial effects of the sanitary and housing measures made by the colonial government. But, generally, the response was to doubt or even reject the figures in preference for making categorical claims about the threat of tuberculosis. In 1926, the government stated that the death rate from tuberculosis had not changed much in the colony.[16] Two years later, it attributed a quarter of the Singapore municipal death rate to tuberculosis.[17] A decade on, while acknowledging that the tuberculosis death rate in the municipality had declined more rapidly than the general death rate, the medical services insisted that the rate itself was 'exceedingly high'.[18] In 1939, Dr A.G.H. Smart warned that the municipality's general death rate of 283 per 1,000 in 1937 was still nearly five times that of 58.4 for England and Wales, although the comparison ignored the urban–rural differences between the countries.[19]

The official and professional notions of the tuberculosis peril growing out of Chinese overcrowding was a strong and continuing one. It originated at the beginning of the twentieth century, riding on the back of the Straits Settlements government's growing concern over Singapore's high mortality rates and in particular, the incidence of tuberculosis at Tan Tock Seng Hospital and the General Hospital in the previous decade. In September 1905, upon the advice of his medical officers, Governor John Anderson wrote to the Secretary of State for the Colonies Alfred Lyttleton, attributing the island's high death rate of over 40 persons per 1,000 to the 'defective conditions of housing' in the town area.[20]

Anderson placed the blame on the shophouses, the common form of housing in the municipality. This area, of only 20 square miles in size, had become severely overcrowded. In the early nineteenth century, it had been the designated European quarter and agricultural land, but was now occupied by large numbers of poor Chinese immigrants. The government had failed to prevent the area from being subdivided into building lots and completely built over with shophouses. The shophouses, built back-to-back, were only several storeys high but over time had been extended backwards so the rooms on each floor were many cubicles deep. These myriad tiny cubicles catered to the needs of Chinese immigrants who could only afford minimal rents but sought to live close to their workplace and community located along specific streets of the town area. In a limited area of space, the shophouses creatively integrated commercial trades on the ground floor and the maximal amount of cubicle housing for workers on the upper storeys.

As James Warren succinctly put it, the arrival of large numbers of mostly single, male sojourners from the south-eastern provinces of China in the last decades of the nineteenth century had transformed Singapore into a 'coolie town'.[21] The shophouses made economic sense for both the Chinese commercial class and coolies; for a long time it was also expedient for the government. Governor Anderson in his letter to Lyttleton noted that the shophouses housed about half of the population of Singapore, who comprised the 'poorest class' of Chinese immigrants but also the essential 'cheap labour' that sustained the prosperity of the entrepôt.[22] The colonial regime had in fact made the land in the town available for housing development. A commentary on the housing crisis in the *Singapore Free Press* blamed the government for having progressively 'sliced up the town lands into the scandalously insanitary lots' in order to fill its coffers.[23] Conversely, the government-appointed Housing Commission of 1918 called the landlords of the shophouses 'blood suckers'.[24]

But by the 1900s, the shophouses had earned the ire of colonial administrators and doctors. Between 1891 and 1911, the municipal population had risen by 24 per cent with only an increase of 15 per cent in the housing.[25] Sanitary and public health concerns came to the fore. Anderson highlighted the lack of airflow in shophouses as a major factor in the spread of tuberculosis. The Municipal Commission also pointed to 'the habit of spitting so universal among the Chinese' as a cause of insanitation, and of tuberculosis and pneumonia;[26] spitting in public places, but not in private areas such as homes, was prohibited under the Anti-Spitting Ordinance. The solution, the governor proposed, would be for the government or the Municipal Commission to take over the 99-year leases of the shophouses and convert them to statutory grants, so that the shophouses could be rebuilt and cleansed block by block. He thought the private sector could carry out the work.[27] But such reforms would be tremendously costly, and it made sense for the government to first appoint a sanitary expert to conduct a survey of the shophouses.

Simpson and the sanitary state of Singapore

The expert nominated was W.J.R. Simpson. He was already a leading British expert on tropical medicine when he came to Singapore in 1906 and wrote the report, *The Sanitary Condition of Singapore* (1907). Born in Glasgow in 1855, Simpson received his Doctor of Medicine at the University of Aberdeen and Diploma in Public Health at the University of Cambridge. He gained his practical experience as Medical Officer of Health of Aberdeen, and then of Calcutta in British India. In 1899, he helped found the renowned London School of Tropical Medicine. Moving along imperial circuits of expertise, Simpson undertook numerous studies of epidemic diseases such as plague, cholera and tuberculosis in British colonies in Asia and Africa. His visit to Singapore followed an earlier one to Hong Kong.[28]

While knowledgeable and highly regarded, Simpson's role in history was nevertheless not merely as an individual. His visit to Singapore tied the island to wider trends in imperial medical history. Simpson belonged to a group of public health experts involved in an expanding collaboration with the British Colonial Office, which actively promoted research on tropical medicine between 1895 and 1914.[29] This collaboration was driven by two developments. One was the emergence of germ theory in the last two decades of the nineteenth century, which largely replaced the miasma theory of disease. Germ theory did not merely trace diseases to pathogens and parasites, but also shaped their study in distinct fields of research and expertise. Two institutions founded in the late 1890s led the way in this research in Britain: the London School of Tropical Medicine, which trained doctors for the colonial service (and where Simpson taught), and the Liverpool School of Tropical Medicine, which provided experts for outbreaks of disease in the tropics. The second factor, which provided the practical motivation for the research, was the growing concern of the Colonial Office in the health of Europeans based in tropical territories.[30]

The crucial point about tropical medicine was that, despite its interest in colonial areas, it essentially derived from metropolitan history and experience. The unmanaged growth of Britain's cities had led to the emergence of slums and sparked fears over the sanitary consequences. Joseph Chamberlain, the Secretary of State for the Colonies between 1895 and 1903, had as mayor of Birmingham waged war on slum housing in the 1870s.[31] British inner cities were also the target of a moral panic among the upper classes in the late nineteenth century.

In London, the environmental deterioration arising from the growth of slums was blamed for rendering the city's casual poor, who dwelt in the inner city, into degenerate 'outcasts' – a threat to the city and moral order.[32] The state of the environment remained central in contemporary medical thinking. While it superseded the miasma theory, sanitary science continued to emphasise the environment as the basis for the growth and spread of germs. To Simpson and his sanitary peers, public health devolved to environmental matters such as the type of urban dwellings, residential density, proximity of buildings, access to air and sunlight, sewage disposal, water supply, climate, and soil.[33] In 1895, the Straits Settlements medical services had linked tuberculosis to the meteorological influences of long, dry periods followed by heavy rains, 'causing abrupt alterations in the level of the sub-soil water'.[34]

These attitudes towards public health can be seen in Simpson's work in cities as varied as Calcutta, Hong Kong, Singapore, and African cities; his advice was consistent. Drawing upon germ theory, he stressed the importance of airflow and light in housing. He also urged the colonial state to take a leading role in improving sanitary conditions, such as by forming a sanitation service.[35] Simpson went beyond many of his contemporaries to underline the health of the natives, as he did in his visits to cities in west and east Africa, and to Singapore. The work of sanitation, in his view, should not be left to locals, who were allegedly ignorant, and the state had to establish building laws and by-laws, and to demolish filthy dwellings.[36]

Arriving in Singapore in May 1906, Simpson spent about three months there. He was predictably concerned with the health of Europeans like his tropical medicine peers. He commented how difficult it was for most Europeans to cope with the debilitating heat and humidity, which was worsened by the poor state of urban sanitation. But Simpson went beyond the Europeans. The general death rate of Singapore was high, rising from 31.27 per 1,000 in 1892 to 43.74 in 1905. The cause was to be found, in his mind, in the other 'races' who populated the settlement, who had failed to adapt to the norms of sanitary urban life:

> the population of Singapore is a peculiar one, in that the immigrants who form such a large population of its inhabitants are mostly people from the rural districts of other countries, and are unaccustomed to the restrictions and practices of town life, and that though the population is quiet and amenable, yet many difficulties are encountered in connection with its sanitary administration that are not met with to the same extent elsewhere.[37]

In particular, Simpson focused on the majority group in Singapore: the Chinese, who were the victims of an 'extraordinary' mortality rate:

> I have had the opportunity of examining the Chinese immigrants as they arrive by ship and it may be stated that, with some exceptions, they represent physically an uncommonly fine body of adults. A high death-rate with such a population means an extraordinary death-rate.[38]

As he discussed the 'principal diseases', his emphasis fell on tuberculosis as the first and most important among them.[39] The disease, he reported, had killed a total of 8,516 people between 1901 and 1905, averaging between 1,700 and 1,852 deaths each year and surpassing beri beri and malaria. In 1905, 17.6 per cent of autopsies performed at TTSH were for tuberculosis cases, just behind beri beri which had the highest ratio (19 per cent). At GH, his report noted, the incidence of tuberculosis had increased nearly three times in the Asian wards between 1893 and 1905, although it had declined in the European wards and was unchanged in the Asian police wards. Simpson added numerous appendices to the report, counting deaths from tuberculosis and other diseases by race and sex in the previous five years. For him, the mortality for tuberculosis was clear: the number of deaths and the death rate were highest among the Chinese and lowest among the Europeans, with Malays, Indians, Eurasians, and 'other nationalities' sandwiched in between. Most of the persons who died from tuberculosis were male, and this was also the case among the Chinese (86 per cent in 1905).

Just as crucially, Simpson looked beyond the hospital for the causes of tuberculosis among the Chinese. His interest was in the environmental aspects of sanitation, and his gaze, like Anderson's, fell on the shophouse. He sent out municipal health inspectors, probably for the first time, to investigate the shophouses and their denizens. His report contained their findings, laid out in tables of statistics, maps, buildings plans, and photographs of the interior of the shophouses along eight streets in the town area, namely, Upper Chin Chew Street, Upper Cross Street, Upper Nankin Street, Cecil Street, North Canal Road, Pekin Street, Hong Kong Street, and Sago Street. The attention to detail showed the great lengths Simpson undertook to study tuberculosis in local housing. He emphasised how the shophouses had been extended back-to-back without open spaces such as courtyards and back lanes. Simpson surmised such construction in decidedly racial terms: 'The Chinese prefer, where possible, to build horizontally rather than vertically.'[40]

The most striking – and also most mistaken – part of the report was a map on page 12, titled 'Deaths from Tuberculosis', which counted the number of deaths in the rooms of two blocks of shophouses along Upper Chin Chew Street and Upper Cross Street for each year from 1901 to 1905. This was an area of 5½ acres, with 194 houses and 492 floors and a population of 3,990 adults and 829 children, or 35.2 houses and 876 persons to an acre.[41] Although this random sample was small, Simpson felt confident that the link between tuberculosis and sanitation had been adequately demonstrated:

> The plan, nevertheless, very graphically illustrates the great incidence of the disease in houses of a class which are numerous in the town.... The disease once in a house tends to recur, some years being worse than others, until it almost assumes an epidemic form.[42]

The spread of the disease in tight spaces meant that, he urged, 'Tuberculosis should be treated as an infectious disease'.[43]

Simpson's report framed an epidemic of the Chinese, dwelling in 'horrific' conditions within the shophouses. He warned that the cubicle housing was 'absolutely destructive to healthy lighting and ventilating of the houses, and to efficient scavenging and drainage of the houses in the blocks'.[44] He highlighted the lack of windows particularly at the back end of the shophouses, where air and light failed to penetrate and it was 'dark and cheerless'.[45] The problem was worsened by the insanitary practices of the residents, one of which was the removal of nightsoil, passing the rooms and corridors of the shophouses. One intimate account in the report of a shophouse along Sago Street told of the inspectors having to use a lamp to find their way about, of hawkers storing live ducks in their room and in the airwell, and of one resident who was ill, coughing and spitting on the floor.

The report also dealt, albeit more briefly, with the sanitary conditions of another racial group – the Malays. Here, Simpson's interest was in wooden kampong housing rather than the shophouse. Simpson surmised that the Malay housing was generally in a better sanitary state than the shophouse, but warned that many Malay wooden houses built on stilts along the banks of the Singapore River had additional smaller dwellings erected in the open space under the house, which were also dark and filthy.

Simpson's focus, though, was undoubtedly on the Chinese denizens in the shophouse. Ignoring the economic factors, he declared the problem to be one of 'haphazard and unhealthy development'. The solution required the 'mapping out of the town'.[46] Simpson emphasised the need for effective municipal planning and controls, as existing building laws and by-laws were inadequate, and for the abolition or improvement of insanitary areas. He compiled a wish-list of proposals for the Municipal Commission: to limit the amount and proportion of housing development in the town, so that the physical area of the building would not exceed two-thirds of the total area of the site; to ensure open spaces for courtyards and backyards, especially for the higher buildings; to provide for back lanes behind and between shophouses; to restrict the depth of a building; to mandate windows or skylights for cubicles; to stipulate the minimum size of a cubicle; and to produce a block plan for street development.

In addition to administrative reforms, Simpson stressed the importance of medical and sanitary expertise. He proposed the creation of a separate sanitary board for the Straits Settlements, which would be chaired by the Principal Civil Medical Officer and would work closely with the Municipal Commission to oversee the abolition of unhealthy dwellings and manage the development of the Malay kampongs. Using the example of the Improvement Trust in Bombay, Simpson highlighted the need for medical representatives on such a board, which the Bombay Trust lacked.

There were unsurprisingly strong accolades for Simpson's study, which, the *Eastern Daily Mail* enthused, 'is likely to remain one of the literary monuments of his untiring zeal and extraordinary hard work in the cause of sanitary science'.[47] But his notions were in general agreement with local views. The 1896 report on the Singapore mental asylum had already attributed the prevalence of tuberculosis to the lack of sanitation, particularly to 'dryness and ventilation'.[48] As Brenda Yeoh observed in her research on the Municipal Commission, local doctors such as Lim Boon Keng and W.R.C. Middleton also viewed tuberculosis as a disease of the urban environment; it was caused by a bacillus, but insanitary conditions made it more virulent.[49] A decade later, medical officer David Galloway held that 'The cubicle question is ... at the root of tuberculosis', but it was 'hopeless' to attempt to educate the Chinese immigrants about the disease.[50] In 1927, J. Tertius Clarke, a long-serving doctor in Singapore and Malaya, concurred that 'Generally speaking the amount of phthisis may be taken to be a gauge of the quantity and quality of house sanitation'.[51]

But Simpson's ideas were also challenged by locals. One newspaper article made reference to earlier proposals by 'our active and energetic [Municipal] Health Officer', [W.R.C.] Middleton, whose recommendations for sanitary reform, made in the Municipal Administration Report of 1905, had been ignored by the Municipal Commission. 'Why, then, it will be asked, have the Government and the Municipality jointly agreed to call in an expert, when the information sought was at hand?', the paper queried.[52] The question was both a critique of the inert colonial administration and its pandering to external experts less informed on local matters. Another newspaper disagreed with Simpson's recommendations on sewage and nightsoil disposal and the water supply,[53] while a Municipal Commissioner in Penang, who was a doctor, rejected the connection between sanitation and the death rates.[54]

The most interesting rejoinder was also the most effective one. A writer in the *Singapore Free Press* doubted the claim drawn from Simpson's map about how tuberculosis was concentrated in shophouse rooms and was a result of congested living. He observed that a large number of 49 rooms had no deaths from tuberculosis at all, compared to just four that had over ten deaths. Why should the concentration of tuberculosis not be consistent throughout the shophouses, given that similar sanitary conditions prevailed? There was, the writer declared, a simpler explanation: that persons seriously ill from tuberculosis were being brought to the rooms to die there, this being a Chinese practice to physically remove them from the community to prevent 'bad joss'. The writer observed that one would not likewise conclude TTSH to be the most insanitary place in Singapore because it had the largest numbers of deaths from tuberculosis! Was Simpson, he asked, not aware of this custom?[55] In his memorandum on tuberculosis submitted to the 1918 Housing Commission, David Galloway agreed that Simpson had been mistaken, and that the shophouse rooms were really the 'houses for the dying' that the Chinese operated throughout the municipality.[56]

Simpson redux: The 1918 Housing Commission

Despite Simpson's pedigree, most of his recommendations were not truly implemented. The colonial government passed his report to a health committee set up by the Municipal Commission, which accepted most of his proposals.[57] A major amendment to the Municipal Ordinance was quickly passed in 1907, containing

Simpson's recommendations to detect overcrowding, remove insanitary houses and control dangerous infectious diseases.[58] But it was easier for the colonial state to pass legislation than to effect real reform. In 1909, the Commissioners demurred on the housing issue, as 'Overcrowding is a very complex question' and merely trying to suppress it would either displace it to another shophouse or drive the tenants into the streets.[59] Although they hailed the key provisions of the amendment for back lanes, open spaces and the reconstruction of unhealthy areas as 'practically an adaptation of the Home Act for the housing of the working classes',[60] the proverbial elephant in the room was where or how to make the coolies' housing clean, yet affordable.

Ominously, the *Straits Times* had warned that Simpson's report was 'voluminous, exhaustive and exhausting, and its recommendations drastic and expensive'.[61] One of the most important parts of the report – the dark, airless cubicles – defied colonial reform. The municipal inspectors tried to pull down the shophouse partitions, but this turned frustratingly into an act of what Brenda Yeoh called Sisyphean labour, as the partitions were 'promptly rebuilt' once the inspectors left.[62] The back lanes were also unpopular, for they sometimes divided families living across the backs of shophouses.[63] In 1911, the colonial medical services were initially hopeful that the sanitary measures would lower the tuberculosis death rate, but soon conceded that it would take time before the impact was manifest.[64]

In 1917, a decade after the Simpson report, a large, follow-up study was undertaken by the Singapore Housing Commission, which was tasked to investigate the overcrowding problem. The Commission revisited many of the issues and themes highlighted by Simpson, including the view of tuberculosis as a disease of urban congestion and the need for public education on hygiene and sanitary matters. Housing, as the name of the Commission implies, was the chief concern. Its report reused the plates of darkly lighted cubicles printed in Simpson's report, and it also contained a memorandum from Middleton, repeating his earlier contention, from 1905, that the mortality of tuberculosis was higher (12 per cent) in shophouses built back-to-back than in those not constructed in this way.[65]

The Commission demarcated the northern and southern halves of the 'Congested Area' of the town, the first being the original Malay quarter lining the Kallang and Rochore rivers, totalling 305 acres in size, and the second being an area of 189 acres which the Commission called Chinatown. By this time, the congested area housed over 310,000 people, many of whom resided in poverty and slum-like conditions. The Commission lamented that this social class 'cares nothing for sanitation, ventilation or even bare comfort'.[66]

Returning to Simpson's call for a sanitary service, the Housing Commission agreed that the problem was now so serious as to warrant the creation of a Singapore Improvement Trust. This would be an agency, distinct from the Municipal Commission, which would carry out improvement schemes and town planning. The Trust would sanitise current housing and establish healthy residential areas. The Housing Commission stressed that it was important to rehouse the shophouse dwellers ahead of demolishing insanitary buildings, so that those evicted would not simply return to the area and worsen the overcrowding.[67] This would become a key principle of housing development for the SIT, discussed in the following chapter, and its postcolonial successor, the Housing and Development Board (Chapter 6).

As significant as the housing proposals was the continuing influence of medical expertise in the Housing Commission. The British doctors who were interviewed by the Commission thought of tuberculosis in racial and statist terms, as Simpson did. J.A.R. Glennie, the acting Municipal Health Officer, believed that 'the Asiatic does not like air in his dwelling-house', and this view was unlikely to change. While he was aware that repatriating tuberculosis sufferers from Singapore was a violation of their rights, Glennie supported 'the principle of it', for 'Singapore is no place for a phthistical patient'.[68] Another doctor, G.A. Finlayson, also called for robust colonial intervention in public health, for 'Viewed from a medical standpoint no individual whatsoever has the slightest right to maintain whatever is shown to be detrimental to the health of his fellow citizens'. He supported strong measures to ensure that the shophouse cubicles receive adequate light and air, for 'The organisms in the sputum of a tuberculosis patient maintain their vitality for a much longer period in a dark dank cubicle than when submitted to the rays of sunlight in an open space'.[69]

The strength of the medical arguments contrasted with the Housing Commission's eventual inaction. The Commission concluded that information on tuberculosis in Singapore was lacking and that an external expert in the mould of Simpson be appointed to study the prevalence of the disease. Nothing came out of this. David Galloway's bold proposal for a health officer to visit and disinfect the deceased's house also came to nought.[70] Of the one recommendation that materialised was tuberculosis being made a notifiable disease in November 1918.[71] But the notification system existed only in name, as most cases continued to be notified only after the sufferer's death – 'the patient never having been seen in life by a medical man'.[72] In 1922, the Municipal Commission reported a disappointing 169 notifications compared to over 1,500 deaths from tuberculosis. As late as 1934, the colonial government stated, notifications of all diseases by private medical practitioners were few as they showed 'great reluctance' to do so.[73] Three years later, the Municipal Commission announced that tuberculosis had 'unexpectedly' been removed from the list of notifiable diseases.[74] It had not been consulted on the matter and lamented that while a toothless measure, notification had at least minor benefits, leading to the cleansing of patients' homes and public education on tuberculosis as an infectious disease.

One other proposal came into effect but it was short-lived: Galloway's call for the establishment of a tuberculosis dispensary, following the Edinburgh system of sanatoria in Scotland.[75] In 1920, the colonial government appointed a tuberculosis officer, W. Dawson, who took charge of a small dispensary set up in Singapore the following year.[76] Despite providing free treatment and circulating publicity pamphlets printed in vernacular languages, the dispensary proved a disappointing failure. It was ignored by the general population and closed down later in the year; it had received only 24 patients in three months, with nearly half of them paying only a single visit.[77]

Singapore's overcrowding did not improve. A Town Improvement bill was prepared, but not accepted, and was quickly overtaken by subsequent developments: a land boom which raised property values, and then a slump that curtailed the government's budget.[78] There was also a lack of available land for new housing, as a large area of the municipality was occupied by cemeteries. In 1921, the medical services stated that overcrowding and sanitation were being dealt with but 'real improvement was not to be expected in the near future'.[79] A letter to the *Singapore Free Press* lamented the following year how nothing had still been done by the government to improve sanitary conditions. It blamed this lack of action on the costs of providing proper services and investigating the major tropical diseases.[80]

The 1923 Tuberculosis Committee

Instead of a major mission headed by an external expert, the colonial government appointed a small Tuberculosis Committee in 1923 to investigate the issues plaguing the 'pestilential cesspool' that Singapore had become.[81] The Committee comprised local doctors in the civil service, some of whom had been consulted by the Housing Commission: A.L. Hoops, the Principal Civil Medical Officer and chairman of the Committee; G.A. Finlayson, the government pathologist; P.S. Hunter, the Municipal Health Officer; W. Dawson, the Deputy Municipal Health Officer. The Committee compiled some statistics on tuberculosis, noting that 1,564 persons had died in the previous year, and that 15 per cent of all certified deaths were from the disease during the period 1913 to 1922.[82] Given the problem of certification, these mortality figures did not constitute comprehensive data.

As with the Housing Commission, the Committee emphasised overcrowding, but it did make more precise comments on dealing with tuberculosis. The Committee had initially only been tasked to consider the treatment of tuberculosis cases, but following an early meeting, it managed to persuade the government that the prevention of the disease was 'infinitely of greater importance'.[83] The *Malaya Tribune* took this as evidence of official disinterest: 'We cannot imagine how anybody could have been so foolish, with the crying need for preventive measures, as to originally ask this Committee to confine its activities to curative measures.'[84]

The Committee duly put forward both social and medical reforms. Of the latter, it underlined housing as the most crucial element, as the population of the town had continued to increase beyond the housing capacity. The Committee asked the government to support improvement schemes as a form of prevention which had yet to be carried out despite the 1907 and 1918 reports. Governor Laurence Guillemard underlined its report's significance:

> The essence of this report is that the provision of sufficient new houses, and the improvement of housing accommodation in the congested areas, must precede or accompany any other action taken to control the spread of tuberculosis and

that treatment of the already infected can of itself have little effect on the spread of this disease, or the resulting death-rate. IN THIS CONNECTION THE IMPROVEMENT TRUST BILL WHICH IS NOW RECEIVING THE CONSIDERATION OF GOVERNMENT IS OF IMPORTANCE.[85]

Specifically, new housing could be built on vacant land adjacent to the congested town area at Cantonment Road and Lavender Street, which would accommodate people displaced by the schemes.

On the negative side of the report, the Committee advocated the repatriation of decrepits and the indigent who could no longer contribute to Singapore's economic interests – something J.A.R. Glennie had proposed to the Housing Commission. It also supported the restriction of immigration on public health grounds.[86] The policy of repatriation had broad support among contemporary European doctors in Singapore. Although not a member of the Committee, David Galloway endorsed such a policy for incipient cases, as 'among immigrant races, a good deal might be done in the way of repatriation', rather than to attempt to treat them in a dispensary. Galloway justified his position on humanitarian and medical grounds:

> This is not advising dumping our undesirables on others but is really a sound policy to follow in the interests of the patient ... the best chance of recovery is by sending them back to an open-air life while yet the disease is young.

For advanced cases, Galloway again proposed the Edinburgh system where they would be segregated from the general population in a sanatorium.[87]

On medical matters, the Committee made submissions on both early and advanced cases. It did not recommend the building of a sanatorium for early cases which would be too costly. The Committee made the rather spurious argument that it did not make sense to temporarily remove cases from the insanitary areas and allow them to return there after treatment. In 1929, Governor Laurence Guillemard gave his support to an ambitious plan to build a sanatorium for all early cases in Malaya, with only the site purportedly to be determined. But nothing came out of the plan, likely due to the effects of the Great Depression later that year on the government's finances.[88]

Instead of a sanatorium, the Committee recommended that 'limited accommodation' be provided for early cases in the newly enlarged premises of GH.[89] This soon increased the number of tuberculosis admissions at the hospital, which, since the 1870s, had not been a primary institution for treating the disease. From 1907 to 1911, there were less than 18 tuberculosis patients per year in the European first and second-class wards, and less than 108 in the Asian wards.[90] But from 1925 to 1933, the number of tuberculosis patients admitted to GH, which was renamed the Singapore General Hospital (SGH) in 1926, rose from 284 to 724 This made it the second most important institution for the treatment of tuberculosis after TTSH, which itself also played a new role as a teaching hospital in the early nineteenth century. Similar small numbers of accommodation for early cases were also made available in the other government hospitals of Singapore and the Straits Settlements, although only SGH offered specialised treatment for the illness. In 1931, these modest measures reportedly had a positive effect in treating the early cases.[91]

The Committee suggested that advanced cases be isolated in a hospital as they formed a 'principal foci of infection' which would pose a danger to the community.[92] Isolation would generally be voluntary, but could be mandated by law if compulsion was needed. The isolation measure showed the influence of the Edinburgh system and local doctors such as Galloway, who endorsed the segregation of advanced cases from the general population.[93] However, in 1929 the colonial government deemed the compulsory isolation of infectives to be 'not yet practicable'.[94]

The Committee also proposed that TTSH accept a small number of advanced cases and treat early cases, but the latter was implemented only half-heartedly.[95] TTSH remained largely a paupers' hospital with no women or children patients with tuberculosis. The absolute numbers of patients treated there jumped from 276 in 1900 to 1,042 in 1938, as did the number of deaths – more modestly – from 186 to 420 over the same period. In 1924, a reported 28 per cent of deaths at TTSH were from tuberculosis.[96] But, in context, from 1927 to 1936 an average of 657 tuberculosis patients were admitted each year to TTSH, which comprised only 5.7 per cent of all admissions.[97] In 1938, there was a reported increase in tuberculosis cases at the hospital, though it was not known if these were advanced or early cases.[98]

The Committee linked the spread of tuberculosis infection to the 'indiscriminate spitting' widely practised by the Asian population. This was occurring not only in the streets but also onto the floors of houses in the congested area.[99] Attempting to deal with this problem brought the colonial government beyond the hospital into public health education. Subsequently, schools and child welfare centres in Singapore began to propagate hygiene matters in a bid to arrest the spread of tuberculosis, along with diseases like malaria, the venereal diseases and hookworm.[100] In 1930, a commentary in the *Medical Journal of Malaya* proposed that the government print campaign posters which had been used successfully in New York, carrying the warning, 'Spitting Spreads Disease'.[101] But the bulk of such health education work had to await the end of the Second World War.

Nonetheless, despite its flaws, Simpson's brief sojourn in Singapore laid the foundation for a tuberculosis control state in the 1920s and 1930s, and particularly after the war. He had relied on local medical expertise and had made serious errors in his study, but his long-term influence on the history of Singapore is significant. All of his recommendations – the notification of cases to the state, the proposal for a sanatorium, the establishment of a dispensary in the community, the use of public propaganda, and the control of shophouses and their coolie-dwellers – formed the basis of a firm anti-tuberculosis programme. In fits and starts, his proposed programme began immediately to take shape among the houses and races of Singapore.

Notes

1 Singapore Municipality, *Administration Report 1905*, p. 54.
2 CO 275/148 1937 Straits Settlements Medical Report, p. 1005.
3 CO 275/70 1904 Straits Settlements Medical Report, p. 706.
4 CO 275/72 Straits Settlements, Registration of Births and Deaths, p. 189.
5 CO 275/72 1905 Straits Settlements Medical Report, p. 682.
6 CO 275/89 1911 Straits Settlements Medical Report, p. 536.
7 CO 275/103 1920 Straits Settlements Medical Report.
8 Singapore Municipality, *Administration Report 1931*.
9 CO 275/100 1918 Straits Settlements Registration of Births and Deaths.
10 Singapore Municipality, *Administration Report 1934*, p. 6-D.
11 Singapore Municipality, *Administration Report 1934*, p. 6-D.
12 CO 275/107 1922 Straits Settlements Registration of Births and Deaths.
13 CO 275/126 1930 Straits Settlements Medical Report.
14 CO 275/148 1937 Straits Settlements Registration of Births and Deaths, p. 1083.
15 CO 275/131 1932 Straits Settlements Medical Report.
16 CO 275/117 1926 Straits Settlements Medical Report.
17 CO 275/121 1928 Straits Settlements Medical Report.
18 CO 275/150 1938 Straits Settlements Medical Report, p. 855.
19 CO 273/655 Memo by A.G.H. Smart, 7 December 1939.
20 CO 273/310 Letter from Governor John Anderson to Secretary of State for the Colonies Alfred Lyttleton, 12 September 1905.
21 James Francis Warren, *Rickshaw Coolie: A People's History* (Singapore: Singapore University Press, 2003).
22 CO 273/310 Letter from Governor John Anderson to Secretary of State for the Colonies Alfred Lyttleton, 12 September 1905.
23 *Singapore Free Press and Mercantile Advertiser (Weekly)*, 27 June 1907.
24 Singapore, *Proceedings and Report of the Commission Appointed to Inquire into the Cause of the Present Housing Difficulties in Singapore, and the Steps Which Should be Taken to Remedy Such Difficulties*, Vol. I (Singapore: Government Printing Office, 1918).
25 Singapore, *Proceedings and Report of the Commission Appointed to Inquire into the Cause of the Present Housing Difficulties in Singapore*, Vol. I, p. A67.
26 Singapore Municipality, *Administration Report 1907*, p. 34.
27 CO 273/310 Letter from Governor John Anderson to Secretary of State for the Colonies Alfred Lyttleton, 12 September 1905.
28 R.A. Baker and R.A. Bayliss, 'William John Ritchie Simpson (1855–1931): Public Health and Tropical Medicine', *Medical History* 31, 1987, pp. 450–465.
29 Joseph Morgan Hodge, *Triumph of the Expert: Agrarian Doctrines of Development and the Legacies of British Colonialism* (Athens, OH: Ohio University Press, 2007), p. 54.
30 Hodge, *Triumph of the Expert*.
31 Hodge, *Triumph of the Expert*.
32 Stedman Gareth Jones, *Outcast London: A Study in the Relationship between Classes in Victorian Society* (Oxford: Clarendon Press, 1971).
33 Brenda S.A. Yeoh, *Contesting Space in Colonial Singapore: Power Relations and the Urban Built Environment*, 2nd edition (Singapore: Singapore University Press, 2003).
34 CO 275/50 Annual Medical Report of the Civil Hospitals in the Straits Settlements for the Year 1895, p. 562.
35 Baker and Bayliss, 'William John Ritchie Simpson'.
36 Hodge, *Triumph of the Expert*.
37 W.J. Simpson, *Report of the Sanitary Condition of Singapore* (London: Waterlow & Sons, 1907), p. 2.
38 Simpson, *Report of the Sanitary Condition of Singapore*, p. 7.

39 The Simpson report also looked at diarrhoea and dysentery and the associated problem of sewage removal and disposal, and at quarantine efforts for smallpox, plague and cholera.

40 Simpson, *Report of the Sanitary Condition of Singapore*, p. 12.

41 Singapore Municipality, *Administration Report 1909*.

42 Simpson, *Report of the Sanitary Condition of Singapore*, p. 10.

43 Simpson, *Report of the Sanitary Condition of Singapore*, p. 10.

44 Simpson, *Report of the Sanitary Condition of Singapore*, p. 12.

45 Simpson, *Report of the Sanitary Condition of Singapore*, p. 14.

46 Simpson, *Report of the Sanitary Condition of Singapore*, p. 29.

47 *Eastern Daily Mail*, 13 June 1907.

48 CO 275/53 Annual Medical Report of the Civil Hospitals in the Straits Settlements for the Year 1896, p. 552.

49 Yeoh, *Contesting Space in Colonial Singapore*.

50 D.J. Galloway, Tenth Meeting, in Singapore, *Proceedings and Report of the Commission Appointed to Inquire into the Cause of the Present Housing Difficulties in Singapore*, Vol. II, pp. B80, B83.

51 J. Tertius Clarke, 'Tuberculosis in the Tropics', *Medical Journal of Malaya* 2 (1), 1927, p. 21.

52 *Eastern Daily Mail*, 12 July 1907.

53 *Singapore Free Press and Mercantile Advertiser*, 13 June 1907, 14 October 1909.

54 *Straits Times*, 14 September 1909.

55 *Singapore Free Press and Mercantile Advertiser*, 27 June 1907.

56 D.J. Galloway, 'Notes on Tuberculosis', 8 November 1917, in Singapore, *Proceedings and Report of the Commission Appointed to Inquire into the Cause of the Present Housing Difficulties in Singapore*, Vol. I, p. C9.

57 CO 275/77 1907 Straits Settlements Medical Report.

58 Singapore Municipality, *Administration Report 1907*.

59 Singapore Municipality, *Administration Report 1909*, p. 39.

60 Singapore Municipality, *Administration Report 1907*, p. 41.

61 *Straits Times*, 12 June 1907.

62 D.J. Galloway, 'Notes on Tuberculosis', 8 November 1917, in Singapore, *Proceedings and Report of the Commission Appointed to Inquire into the Cause of the Present Housing Difficulties in Singapore*, Vol. I, p. C9; Brenda S.A. Yeoh, *Contesting Space in Colonial Singapore* (Singapore: Singapore University Press, 2003).

63 Seah Liang Seah, Fifteenth Meeting, in Singapore, *Proceedings and Report of the Commission Appointed to Inquire into the Cause of the Present Housing Difficulties in Singapore*, Vol. II.

64 CO 275/87 1911 Straits Settlements Medical Report; CO 275/89 1912 Straits Settlements Medical Report.

65 Major W.R.C. Middleton, Memorandum in Singapore, *Proceedings and Report of the Commission Appointed to Inquire into the Cause of the Present Housing Difficulties in Singapore*, Vol. I.

66 Singapore, *Proceedings and Report of the Commission Appointed to Inquire into the Cause of the Present Housing Difficulties in Singapore*, Vol. I, p. A6.

67 Singapore, *Proceedings and Report of the Commission Appointed to Inquire into the Cause of the Present Housing Difficulties in Singapore*, Vol. I.

68 J.A.R. Glennie, Memorandum in Singapore, *Proceedings and Report of the Commission Appointed to Inquire into the Cause of the Present Housing Difficulties in Singapore*, Vol. I, p. C17; Thirteenth Meeting, in Singapore, *Proceedings and Report of the Commission Appointed to Inquire into the Cause of the Present Housing Difficulties in Singapore*, Vol. II, p. B119.

69 G.A. Finlayson, Memorandum in Singapore, *Proceedings and Report of the Commission Appointed to Inquire into the Cause of the Present Housing Difficulties in Singapore*, Vol. I, p. C111.

70 David J. Galloway, 'Notes on Some Local Aspects of Tuberculosis', *Medical Journal of Malaya* 3 (1), 1928.
71 G.A. Finlayson, Memorandum in Singapore, *Proceedings and Report of the Commission Appointed to Inquire into the Cause of the Present Housing Difficulties in Singapore*, Vol. I; J.A.R. Glennie, Thirteenth Meeting, in Singapore, *Proceedings and Report of the Commission Appointed to Inquire into the Cause of the Present Housing Difficulties in Singapore*, Vol. II; Lim Boon Keng, Seventeenth Meeting, in Singapore, *Proceedings and Report of the Commission Appointed to Inquire into the Cause of the Present Housing Difficulties in Singapore*, Vol. II.
72 Singapore Municipality, *Administration Report 1924*, p. 4-D.
73 CO 275/136 1934 Straits Settlements Medical Report, p. 999.
74 Singapore Municipality, *Administration Report 1937*, p. D-1.
75 David J. Galloway, 'Notes on Some Local Aspects of Tuberculosis', *Medical Journal of Malaya* 3 (1), 1928.
76 SIT 662/31 Memo from W. Dawson to Municipal Health Officer; D.J. Galloway, Tenth Meeting, in Singapore, *Proceedings and Report of the Commission Appointed to Inquire into the Cause of the Present Housing Difficulties in Singapore*, Vol. II.
77 D.W.G. Faris, 'Some Figures Relating to Tuberculosis in the Straits Settlements', *Journal of the Malayan Branch of the British Medical Association* 1 (3), December 1937, pp. 211–216.
78 CO 273/529 Memo by P.S. Hunter, 3 July 1925.
79 CO 273/518 Notes by the Medical Secretary on the Annual Report of the Medical Department, 1921, p. 18.
80 *Singapore Free Press and Mercantile Advertiser*, 24 August 1922.
81 *Straits Times*, 3 November 1923.
82 *Malaya Tribune*, 7 November 1923.
83 *Malaya Tribune*, 7 November 1923.
84 *Malaya Tribune*, 7 November 1923.
85 Singapore Municipality, *Administration Report 1923*, p. 3, original emphasis.
86 *Straits Times*, 3 November 1923.
87 David J. Galloway, 'Notes on Some Local Aspects of Tuberculosis', *Medical Journal of Malaya* 3 (1), 1928, p. 4.
88 CO 275/124 1929 Straits Settlements Medical Report.
89 *Straits Times*, 3 November 1923.
90 CO 275/87 1911 Straits Settlements Medical Report.
91 CO 275/128 Memorandum on the Incidence of Tuberculosis in the Straits Settlements, 31 August 1931.
92 *Straits Times*, 3 November 1923.
93 David J. Galloway, 'Notes on Some Local Aspects of Tuberculosis', *Medical Journal of Malaya* 3 (1), 1928, p. 4.
94 CO 275/124 1929 Straits Settlements Medical Report, p. 653.
95 W.E. Hutchinson, 'Some Aspects of the Tuberculosis Problem in Singapore', *Journal of the Malayan Branch of the British Medical Association* 1 (3), December 1937, pp. 218–229.
96 CO 275/112 1924 Straits Settlements Medical Report.
97 R.A. Pallister, 'Some Observations on Pulmonary Tuberculosis in Singapore', *Journal of the Malayan Branch of the British Medical Association* 1 (3), December 1937, pp. 231–235.
98 CO 275/150 1938 Annual Report of the Medical Officer-in-Charge of Tan Tock Seng Hospital for the Year 1938.
99 *Singapore Free Press and Mercantile Advertiser*, 31 October 1923.
100 CO 275/124 1929 Straits Settlements Medical Report.
101 'A Campaign against Tuberculosis', *Medical Journal of Malaya* Vol. IV, 1930, p. 73.

3   Houses and races of the colony

The influence of W.J.R. Simpson's 1907 report and follow-up studies on housing and tuberculosis became manifest in the 1920s and 1930s. Their recommendations on sanitation and housing formed the basis for an anti-tuberculosis policy, while also having the bigger effect of shaping Singapore into a governable colony in both physical and social terms. The British colonial government thus began to imagine and implement a pair of policy projects from the earlier recommendations: the first was to transform the physical environment of the town, while the second was to reform the unruly Chinese coolies who dwelt there.

The former would lead to the formation of the Singapore Improvement Trust (SIT) in 1927, a specialist agency that was independent of the Municipal Commission and informed by medical expertise. The Trust quickly moved to build houses for the working class to replace the overcrowded and squalid shophouses in the town area in the early 1930s. In 1934, the medical services of the Straits Settlements lauded the building of modern housing which would replace the slums in Singapore as 'one of the most potent anti-tuberculosis measures we have'.[1] This view of tuberculosis control heralded the origins of public housing in Singapore.

But alongside the will to rebuild the urban environment was the colonial doctors' interpretation of sanitation in pseudo-scientific ways – in terms of governable racial groups. This perspective stemmed from the prevailing medical theory that attributed tuberculosis to the purported biological immunity and resistance, or lack thereof, among the various races. Tuberculosis was a social illness: it signified the Chinese not only as people who lived and died in filthy spaces, but also as a distinct race possessing a certain amount of immunity and resistance to disease. The Chinese, in this theory, were relatively more immune and resistant than the other ethnic groups in Singapore. In 1921, the government noted, 'The hospital death rate of nearly 50% [in the Straits Settlements] indicates the low resistance offered by Eastern races'.[2] The idea of racial immunity and resistance expressed the sociocultural views of tuberculosis among colonial (and sometimes even Asian) doctors and officials. Its central figure was the poor, urban-dwelling Chinese coolie, against whom both the Europeans and other Asian ethnic groups were to be contrasted and defined.

Belatedly, then, tuberculosis gained a high place as a killing disease in 1930s Singapore. In 1929, the British administration noted that the urban death rate in the Straits Settlements was more than double the rural one.[3] Up to the outbreak of the Second World War in Malaya in 1941, much talk was made and ink spilt over the disease across the pages of government reports and medical journals by Western (and a few Asian) physicians and administrators, both municipal and colonial. Not much was accomplished besides the limited SIT housing, whether to combat the disease's spread, to improve sanitary conditions in the town area, or to reform the habits of the urban dwellers. But the ideas, frameworks, debates, experiments, and even failures in this period were indicative of a maturing regime of colonial rule in Singapore. In the longer run, they would shape systems governing housing, public health and ethnicity.

Origins of public housing: the Singapore Improvement Trust

By the time of the 1923 Tuberculosis Committee, there was growing public demand for action. In response to the Committee's report, as noted in the previous chapter, Governor Laurence Guillemard emphasised how any anti-tuberculosis work must be supported by 'the provision of sufficient new houses, and the improvement of housing accommodation in the congested areas'.[4] The Municipal Commission concurred: 'the matter is in our hands, and is not a matter for speculation'.[5] In 1926, it advised that the new houses to be built should not exceed three storeys (a common height of the shophouses), otherwise they would degenerate into a new type of slum.[6] But the colonial officials were not really espousing a new approach to an old problem, for such views had become widespread, growing in urgency and the need for immediate action.

In the same year as the report, a commentary on slum housing in the *Malaya Tribune* warned that 'no less than 1,644 deaths from phthisis and tuberculosis were recorded in Singapore, and that the tuberculosis death rate of Singapore is the highest known in the world'.[7] The commentary underscored a sense of déjà vu: the Tuberculosis Committee's ideas were hardly novel, as Middleton had anticipated them back in 1905.[8] In 1925, concern with what had become known in the public sphere as the 'terrible slums' of Singapore reached a high point.[9] In April of that year, another *Malaya Tribune* article warned about the dire consequences of the spread of cholera and tuberculosis.[10] In August, the *Straits Times* blamed Guillemard for failing to act on the recommendations of the Tuberculosis Committee despite all the scientific evidence presented, notably from the government's own doctors.[11]

The following month, a seemingly minor but politically significant controversy erupted when James Lornie, the outgoing Deputy President of the Municipal Commission, opposed the establishment of the proposed Improvement Trust. Instead, he proposed the setting up of a committee comprised of two Municipal Commissioners and a Legislative Councillor, which would undertake to 'clean up' Chinatown by building new back lanes and open spaces. It was too difficult, he thought, to rebuild the entire congested area.[12]

His real reasons were, it seems, to keep the powers of the new agency within the municipality and to limit the work to improvement rather than redevelopment. Lornie had been a member of the 1918 Housing Commission, where he had expressly opposed the creation of the Trust, preferring the improvement work to remain with the Municipal Commission.[13] But the other Municipal Commissioners objected to Lornie's plan in a subsequent meeting in October, reiterating the need for decisive government action on the slums. The new President of the Commission R.J. Farrer, referring to the 1907 and 1918 reports, supported the formation of the SIT. He asked for the Trust to be endowed with adequate executive powers and funds to carry out its work; tuberculosis was, he warned, 'our worst curse'.[14]

Farrer's opinion won out. In 1927, after a delay of two decades from the Simpson report, the SIT was finally formed. Its task was to 'provide for the Improvement of the Town and Island of Singapore', for which it received a government fund of $10 million to tackle the slum problem. By this time, overcrowding in the town due to the increase in Singapore's population was as bad, if not worse, than in 1905 when W.J.R. Simpson had arrived.[15] On the board of the Trust were the President of the Municipal Commission, who was also the Chairman of the Trust, and the Municipal Health Officer, both of whom were ex-officio members.[16] This was the medical expertise that Simpson had sought for a sanitary service, which enjoined the programmes of sanitation, tuberculosis response and housing.

The Trust's early work was shadowed by the spectre of tuberculosis infection highlighted by the reports of 1907, 1918 and 1923. It initially had an improvement mandate, being tasked to acquire land to carry out improvement schemes, build back lanes behind shophouses and condemn insanitary buildings. The Trust determined that no new housing be built without back lanes.[17] Within several years, it was reported that the improvement schemes were having a positive impact in reducing overcrowding and poor ventilation in the shophouses, so much so that the medical services suggested it was time to revisit the idea of establishing tuberculosis clinics in the town.[18] The Municipal Commission was, however, more pessimistic, deeming the Trust to have made 'little impression' on the slums.[19] In 1936, sanitary inspectors made over 10,000 house to house inspections in Singapore, although what transpired in improving the housing conditions is unknown.[20] In 1938, however, the Weisberg Committee was established to oversee the redevelopment of Crown lease land in the town. It reported that, with the doubling of the urban population in the last five years, the shophouses 'have become warrens of cubicles and form slums of the worst description, and are a proved menace to the health of the people'.[21] Likewise, the Commission reported the following year that the population density in five severely congested blocks of shophouses had risen since 1933.[22]

Housing development was not its primary work but the Trust soon found itself having to build accommodation to rehouse those displaced by its improvement schemes. The Trust was quickly forced to abandon the policy of condemning insanitary houses, such as those at Sago Street and Smith Street which were even more overcrowded in the early 1930s. It was too expensive to acquire and demolish the shophouses. Instead the Trust began to build new housing close to the congested areas, as proposed by the 1923 Tuberculosis Committee.[23] Thus the Trust commenced the development in 1936 of a housing estate in Tiong Bahru, after a devastating fire two years earlier had cleared the squatter settlement there. The estate, it was hoped, was close enough to Chinatown to appeal to its masses. In 1941, on the eve of the Pacific War, the Trust had built 784 flats, 54 tenements and 33 shops in Tiong Bahru. The Straits Settlements medical services recognised the Trust's efforts, but felt that progress was 'very slow'.[24]

Constructing the population

While the SIT commenced its long-term housing work, the government's medical doctors engaged a puzzling question over tuberculosis. In 1937, the Acting Chief Health Officer D.W.G. Faris pointed to a significant decline in the incidence of tuberculosis and pneumonia combined (since the former was often mistaken for the latter) from 730 per 100,000 in 1927 to 517 in 1936. Similarly, the proportion of deaths attributed to tuberculosis of all diseases had dropped slightly from 17 per cent in 1907 to 13.99 per cent for the period 1907 to 1916, to 12.54 per cent for 1917 to 1926, and more dramatically to 8.7 per cent for 1927 to 1936.[25] Faris attributed the declines partly to the work of the SIT, and partly to the repatriation of tuberculosis patients to their home countries. He proposed that the government implement a version of the tuberculosis villages in the Papworth Home Scheme in England, which would allow early cases to be removed from densely populated areas and isolated in a village settlement with their families.[26] Faris presented his findings to a meeting of British doctors in September that year, which, along with their contributions, were published in a substantial special issue of the *Journal of the Malayan Branch of the British Medical Association* on tuberculosis in Singapore and the Straits Settlements.

Ostensibly a debate over figures and movements, the discussion conjoined in the meeting and the special issue reflected the doctors' broader thoughts and deeper concerns on the population and society of colonial Singapore.[27] While discerning and appraising Singapore's demographic and social trends, the doctors attempted to mentally imagine and construct the population. They raised issues which would be revisited more substantially after the war and which would be key to the development of the tuberculosis control state.

There was an acute awareness, as B.M. Jones explained in his introduction to the issue, of the connection between tuberculosis and Singapore's large 'floating population', but also of the implications of the island's transition to a more settled society.[28] The awareness was a response to the social and demographic trends of the inter-war years, particularly the immigration trends. In the mid-1920s, the Straits Settlements medical services had linked the rise in cases of tuberculosis and pneumonia in Singapore and Penang to worsening overcrowding in the two cities. This was due, they speculated, to the influx of immigrants from China, who were fleeing the political instability there, and also responding to the boom in rubber and tin in Malaya.[29] In 1929, the medical department felt that stricter control of immigration was 'very desirable' from a public health perspective.[30]

In the early 1930s, when the Great Depression decimated the trade of the Straits Settlements, the colonial government repatriated all jobless Indians and smaller numbers of destitute and decrepit male Chinese to their home countries. It also reduced the quota for male Chinese entering Singapore.[31] These were temporary responses to the economic conditions, but Chinese women were allowed to enter until 1938, which had the effect of doubling the sex ratio from the 1911 level. This dramatically increased family life among the population, but as the municipal population jumped from 360,000 in 1923 to 520,000 in 1937, its other consequence was to worsen urban overcrowding in Singapore. This had led the government to task the SIT to build houses for lower-income families to relieve the inner city congestion.[32]

In this context, the doctors looked beyond the statistical declines to highlight the future dangers posed by tuberculosis. One theme in the special issue was migration and family life. There was the old threat posed by immigrants bringing the infection with them into overcrowded shophouses, but this was now being complicated by the emergence of new families with infants and children. In the lead article, the Acting Chief Medical Officer of Singapore D.M. McSwan warned that the real incidence of tuberculosis in Malaya and Singapore was greater than the official figures provided by the hospitals. He linked tuberculosis to the 'semi-migrating immigrant population, living under semi-civilised conditions'. Only one-fifth of the tuberculosis patients admitted to hospitals, he noted, were born in Malaya. The transient majority amounted to half of the Chinese population and two-thirds of the Indians, but their numbers fluctuated with the economic conditions. As McSwan stressed, it was unknown how many of the migrants entered Malaya and Singapore already infected with tuberculosis and how many departed after being infected locally. He believed the disease to be most prevalent among adults between the ages of 21 and 45 inclusive, while being relatively rare among infants and young children. Malaya and Singapore may, he concluded, soon face the difficulty of building large, expensive accommodation for tuberculosis patients, as the government had to do for sufferers of leprosy and mental illness.[33]

Agreeing with McSwan, R.A. Pallister from the College of Medicine in Singapore felt that tuberculosis had become widespread but distinguished between the permanent and transient sections of the populations. He argued that infection among the former typically occurred in children, but in the latter among adult migrants to Singapore. Pallister drew these conclusions from his study of 2,000 Chinese admissions to TTSH in the past four years, where the biggest group were from the ages of 30 to 40, which he deemed to be the infection of newly arrived immigrants. The next largest group were those 45 to 50 years of age and permanently settled in Singapore, who were infected in their childhood and developed tuberculosis later when their immunity failed.[34] Another contributor G. Haridas examined the Singapore General Hospital's records of tuberculosis among children and infants. He admitted that the hospital did not treat many cases of tuberculosis, but extrapolated the dangers of 'familial contagion' in 'tuberculosis environments'. The risk was especially great for infants and children of poor economic backgrounds, who lived among families with 'defective sanitary habits and customs'.[35]

The other theme in the doctors' discussions was, naturally enough, the housing problem. Another contributor to the issue, the Assistant Health Officer of Singapore Municipality W.E. Hutchinson, insisted that tuberculosis was still a serious problem, repeating the adage that the government had failed to implement Simpson's proposals. The key to tuberculosis control was not to provide specialised medical services such as sanatoria, but to use 'general weapons', in particular to reform 'hygiene and social standards'. The fall in the tuberculosis death rate, Hutchinson urged, was remarkable, given the large numbers of migrants arriving in Singapore. He considered that 'it is not surprising to find that phthisis and pneumonia exact their toll when the immigrants are crowded into conditions such as we have here today'. Whatever the figures suggested, Hutchinson placed greater emphasis on the problem of overcrowding. He produced other statistics and building plans illustrating the congestion in the shophouses and urban kampongs. He envisaged an ambitious response: the SIT would build affordable housing for the working class and the proliferation of unplanned huts had to be controlled. A junior officer at the time, Hutchinson would play an important role after the war in the Trust's housing programme and campaign against urban squatter housing.[36]

Chinese immunity and resistance

It was clear from the doctors' discussion above that tuberculosis was not merely a serious infectious disease. It was a social construct which was connected to wider housing, migration, demographic, and social developments. In particular, the disease was a cultural artifice which consolidated and expressed colonial beliefs about the non-Western 'races' and working classes. In this sense, even though the official measures and debates in the early twentieth century had relatively little effect on tuberculosis, the social and cultural impact was much greater. The colonial and medical responses perpetuated racial and class divisions between the ruling elite and the Asian society. Even within the Chinese population, as we will see, tuberculosis reinforced divides between the socio-economic elite and the majority coolie class.

The prevalence of tuberculosis among the Chinese of Singapore was central to the state's imaginary of the people it governed. Racialised ideas of tuberculosis, as we have seen with Simpson, were common among the doctors. Trying to explain Singapore's high death rate in 1907, David Galloway surmised that unlike 'a European populace willing to assist the sanitary authorities', no such assistance could be expected from the Chinese. He added emphatically, 'I refer solely to the thorough Chinese. Their intelligence is so be-fogged by ideas of personal health which have been handed down through countless generations as to become a hereditary instinct'. He deemed cities in China as instances of 'filth', before concluding that 'The Chinaman is the most insanitary of mortals'.[37]

But it was not only gross, misinformed racism that surfaced. Racist views were buttressed by contemporary medical thinking. Lodged in the report of the 1918 Housing Commission was a note from Galloway on the comparative resistance to tuberculosis of various Asian races. It declared that among Malays and Tamils, tuberculosis was rare ('wonderfully rare' among the latter though increasing among the former). For Portuguese and Eurasians, pulmonary tuberculosis was prevalent, especially among the former, and it had been 'exterminating' families. Tuberculosis was common among the Straits Chinese but they had built up resistance to it. The females were typically infected in their homes but males by 'some outside immoral influence'. The China-born Chinese also possessed a 'wonderful degree of resistance'; however, members of the large coolie class 'almost invariably' died when they contracted the disease. For the Japanese, whose numbers in Singapore were becoming significant at the time, tuberculosis was common and also 'very fatal', for they lacked resistance.[38]

In a later article published in 1928, Galloway systematically ranked the races in terms of their 'relative unprotectedness' to tuberculosis: 'Malays, Japanese, agrarian Chinese, some Indian races, notably Sikhs'. He held that 'The phase of civilisation in which most of the races under consideration are, lends itself to the rapid spread of tuberculosis and to the establishment of a rapidly fatal type of disease'.[39]

Galloway's remarks were not individual prejudices but derived from pseudo-scientific notions of racial immunity and resistance that were widely held in metropolitan medical thinking on tuberculosis in the early twentieth century. Michael Worboys calls these notions a 'synthesis of ideas of race, evolutionary theories and immunology'.[40] As he argues, such epidemiological and pathological interpretations of tuberculosis helped shape racial worldviews in North America, Britain and in British colonies. Ideas of immunity and resistance in relation to tuberculosis withstood the general trend in contemporary medical sciences that had begun to question the concept of race. The ideas were compelling because they appeared to explain the incidence and mortality of tuberculosis in both singular and comparative contexts. Ambivalent and contradictory, they were ideologically expedient: they could account for similar trends between white and non-white populations by underlining the influence of environmental, social and cultural factors, but while retaining the category of race. Conversely, the notion of racial susceptibility also excused mining companies in South Africa of their culpability for the high incidence of tuberculosis among their workers.[41]

Similarly, in 1927, J. Tertius Clarke wrote in the *Medical Journal of Malaya* that tuberculosis was a 'bed-room disease', since infection was most likely to occur at home. However, Clarke pointed out that tuberculosis had decreased in England due to improvements in housing, sanitation, people's habits, and segregating the sick, which helped build 'a large amount of immunity – usually acquired in most European people'.[42] D.M. McSwan, writing of the situation in the Federated Malay States, declared that 'repeated implantation of the seed [the disease] will overcome the most resistant constitution'.[43]

Such views prevailed generally in the colonial medical services of the Straits Settlements. They reported in 1924 that tuberculosis was 'very prevalent' among Asians, which was 'a population with a low resistance, as in many cases the fatal termination is a matter of a few weeks only'.[44] Seven years later, rather than acknowledging the social and economic factors, the colonial doctors lamented that

> large numbers of the Asiatic population have not yet acquired a sufficiently high degree of immunity to the disease to escape infection, and that they lack the vital resistance which is necessary to avoid the ravages of the disease when it is once established.[45]

Ideas of racial immunity and resistance shaped colonial policy and discourse in several ways: it absolved the government and the doctors of their failure, or unwillingness, to provide adequate services, while it also distinguished Asians as races who were biologically inferior to Europeans, and who were doubly ignorant about their own bodies and health.

Cultural imaginaries of tuberculosis did not exist only among European doctors. An interesting example is Lim Boon Keng, a medical doctor of Peranakan descent who had been a member of the Straits Settlements Legislative Council. At the time of Simpson's visit, he had ten years of experience in treating tuberculosis. In a 1904 article published in the *Journal of the Malayan Branch of the British Medical Association*, he ascertained tuberculosis to be a common disease 'affecting nearly all classes of the Chinese community'.[46] Lim accepted tuberculosis to be an infectious disease, detailing a case of how members of a rich, educated Chinese family in Singapore had in turn fallen sick with it. He concurred with the British doctors that the disease was most prevalent among patients aged between 15 and 35 years and relatively uncommon among the elderly.

Lim was also a member of the Municipal Commission and like Simpson, linked the spread of tuberculosis to the prevalent lack of sanitation in numerous roads, houses, theatres, coolie houses, and brothels in Singapore, as well as the common habit of spitting on the ground. Urging the colonial government to play a bigger role, he made two proposals: that tuberculosis be made a notifiable disease, and that the government set up a special hospital for tuberculosis patients, so that the disease would not spread to other patients. This would not be a sanatorium, but a hospital for well-to-do patients who were able to pay for their treatment.[47] He later submitted the proposal to the 1918 Housing Commission, relating that he had tried to establish a small hospital at Pearl's Hill, which had failed. But Lim objected to compulsory tuberculin treatment.[48]

However, Lim also had syncretic views of tuberculosis that mixed notions of race, class and Chinese customary beliefs. In the *Journal of the Malayan Branch of the British Medical Association* article, he subscribed to a distinct category of

Chinese tuberculosis and believed the disease to be partly hereditary. His view, and plausibly main experience, of tuberculosis was strongly middle class:

> Many Chinese with hereditary disposition to phthisis have the combination of the following features: – smooth, glossy, delicate skin, beautifully formed face with clean cut features, and finely shaped fingers tapering to sharp points. The eyes have delicate lashes, but this is not as constant as the other features.[49]

Lim's account also shows how far customary Chinese views of tuberculosis differed from British perspectives. Lim opposed the building of a sanatorium to isolate patients, as it was more important to educate the population and help them understand the disease. He described a view of tuberculosis common among the Chinese, one that was striking in appearing to combine traditional and modern ideas:

> The Chinese consider phthisis to be a disease caused by 'living germs,' which leave the patient in swarms as soon as death supervenes, in order to seek pastures new in the shape of other human beings. The people, there-fore, dread very much approaching persons dying from phthisis. They have a superstition that these germs have a predilection for omelette, and that they will fly away, if direct sunlight is thrown into the house by making an opening on the roof. Consequently, it is the custom to place a broad piece of omelette on the face of the dead and to make an opening in the roof, as soon as a phthisical person dies.[50]

Besides Lim, another fascinating account of tuberculosis comes from Chen Su Lan, a medical doctor and social reformer who was a leading advocate of the campaign against opium addiction. Writing in the *Medical Journal of Malaya* in 1932, Chen estimated that there were some 35,000 tuberculosis sufferers in Singapore, or one out of every 12 persons in the municipality, with the disease most prevalent among the Chinese. He dismissed the published statistics based on notifications of tuberculosis, which had risen by nearly 200 per cent from 1924 to 1930 and which pointed to a decline, as in his view many cases went unnotified, unknown or misdiagnosed. He also largely disregarded the SIT's ongoing efforts to build back lanes, roads, sewers, and sanitary houses. Instead, he thought the menace of tuberculosis remained great, which he attributed to the habit of opium smoking among the Chinese. For instance, the germ could be transmitted to addicts through the opium pipe. Chen provided no evidence for his hypothesis. But what was interesting was his damning assessment of the habits of his fellow Chinese, namely, their poor dental hygiene, common-place spitting and even laziness and apathy towards hygiene, all contributing to the spread of tuberculosis.[51]

Chen's article was, however, criticised two years later by a British doctor, J.W. Winchester, who adjudged the notification statistics Chen cited for a period of eight years to be 'valueless' as a gauge of the incidence of tuberculosis. It was only that notification procedures had become more efficient in this short period. Instead Winchester tried to map the disease over 30 years. He found that deaths from tuberculosis in the Straits Settlements had fallen by half from 425.3 per 1,000 persons in 1904 to 208.8 in 1933, and in Singapore, which had the smallest rural population in the colony, even more sharply from 732.67 to 256.06 per 1,000. The death rate for tuberculosis, too, had fallen far more than the general death rate.

Winchester concluded that the reason for the decline was rapid urbanisation, which had swiftly killed off the 'weakly immune'. Disagreeing with Chen, he held that housing schemes would not improve the situation; in fact, a successful housing programme would leave the population susceptible to tuberculosis. He cited historical examples of the vulnerable American Indian population and the black labour troops in French Africa, both of whose numbers were decimated by tuberculosis. It was necessary, he concluded, to raise people's immunity through 'inoculation of the living bacillus', in reference to the work of the French physician Albert Calmette, who had helped discover the Bacillus Calmette-Guérin, or BCG, which was first used as a vaccine in 1921.[52] The debate between Chen and Winchester revealed cleavages between the different approaches to tuberculosis: the improvement of housing and sanitation and the use of inoculation.

Fear of infection and stigma

Outside of the small circle of doctors, in the shophouses where the coolies lived, social perception of tuberculosis had turned into a growing fear of a dangerous infectious disease. As in the nineteenth century, this continued to feed into the persistent reluctance of working-class Chinese to seek hospital treatment. The reluctance was still due to the same reasons as before: the fear of hospitals as death places and the socio-economic repercussions of not being able to work while being treated. But, in addition, there was now increasing social stigma towards people infected with tuberculosis. The discrepancy between the total numbers of tuberculosis deaths in Singapore and in Singapore's hospitals for which we have statistics was marked. In 1904, 499 deaths were recorded at local hospitals and asylums, out of a total of 1,644 for the whole island, or just over 30 per cent. In 1913, the proportion was 36.5 per cent.

In 1908, the management of TTSH reported that 'the majority of these cases [of tuberculosis] are hopeless on admission'.[53] Absconsion of all patients remained a problem at the hospital, and in 1910, the management considered erecting a fence around the premises to deter patients from leaving or bringing in 'undesirable food', as well as unwanted visitors.[54] In 1937, the Straits Settlements medical services were optimistic that 'more patients are presenting themselves for treatment at a comparatively early stage, when treatment by modern methods such as artificial pneumothorax is hopeful'.[55] The following year, the colonial regime would allow government officers to have at least six months' free treatment for tuberculosis in a hospital, which would help them cope with the financial aspects of treatment. There was also a 'noticeable increase' in early cases at TTSH who were better educated persons and thus had a reasonable chance of recovery, although most admissions were still of poor individuals at an advanced stage of the illness.[56] But generally, most coolies of Singapore still did not visit the hospitals willingly: 'This reluctance is forced upon them by the necessity in most cases of providing for the daily needs of their families'.[57]

The public response to tuberculosis revealed the extent of the stigma. If Singapore's working class shunned the colonial hospital, the educated Asian elite were not so ignorant of the basis of modern medicine, although their understanding was also incomplete and flawed. Plans to build tuberculosis wards or sanatoria in the early twentieth century provoked fears among local residents, especially those of the upper classes. As early as 1914, there was an official proposal to utilise the vacated mental illness wards in Pasir Panjang hospital on the southern sea coast of Singapore to treat advanced tuberculosis cases, in order to prevent the spread of infection in the town, but this did not materialise.[58]

In 1927, strong opposition arose to news of government plans to build a sanatorium at a vacant site, also in Pasir Panjang and a popular recreational area.[59] The opposition forced the government to abandon the project and instead consider building the sanatorium at the Trafalgar rubber estate in the distant northeast of Singapore, where new premises for the leprosarium and mental hospital were being planned. Chinese businessman Tan Soo Bin, the owner of Dingwall house close to the proposed site, conveyed a strong objection to the government which is interesting for its mix of science and stigma. He warned of 'not only a fear, but a great risk of infection from the following sources': namely, 'from the air – dust and droplets from sputum' and from

> sewage contamination of the water, as it will be difficult, if not impossible, bearing in mind the class of patients who will be accommodated there, to prevent such patients spitting, urinating and possibly defecating at the sea-front, particularly so at night: – and even if this could be prevented, unless the sewage system were (*sic*) very good, there would still be risk of infection to the residents.

Tan added that 'if the system of isolation is not good, the patients may contaminate the foodstuffs at the shops in the neighbourhood, or those carried by hawkers'. More dubiously, he pointed at the alleged risks from flies – 'flies, which are a pest at certain seasons of the year, may also carry infection to foodstuffs, both raw and cooked, in the neighbourhood' – and from tuberculosis patients swimming in the sea. Tan wanted the sanatorium to be sited in an isolated area away from the general population.[60] His protest against the planned sanatorium was not only effective, but also historically interesting. It showed the cultural effects of a partial understanding of infection in reproducing social stigma against tuberculosis.

By the time the world war reached Malaya in 1941, the British officials and doctors of the Straits Settlements had established a historically conditioned approach and language for governing the colony as well as controlling tuberculosis.

They deemed the crux of the issue to be both the urban houses and the Chinese coolies who dwelt in them. The disease was accepted to be a problem of urban overcrowding and the lack of sanitation. But it was also attributed, somewhat confusingly, to the immigration of working-class Chinese to the town area, to their cultural practices and habits, and to their weak natural immunity and resistance. Besides the doctors, tuberculosis also preoccupied the minds of municipal and housing officials, who were dialogically influenced by the doctors' notions of racial immunity and resistance. Geographically, while Singapore and Penang were deemed to be of foremost concern, by the end of the 1930s official interest in tuberculosis had extended to the rest of British Malaya, such as its spread to the housing of another Chinese coolie group – tin miners – in the Federated Malay States. In 1939, A.G.H. Smart warned the colonial government that 'Malaya is well ahead in most measures for the public welfare, but in respect of tuberculosis seems to have lagged behind'.[61]

Smart's remarks highlight the lack of real action and accomplishment in tuberculosis control efforts. Three decades following W.J.R. Simpson's visit to Singapore, the medical services were still awaiting a comprehensive statistical study on the incidence of tuberculosis among the general population; they had only vague, often unreliable figures on the mortality. They were sufficiently frustrated to report in 1938, 'The progressive increase in the number of admissions to hospital is again apparent. These figures have no relation to the incidence of the disease'.[62] The majority of Chinese, about whom the doctors and officials spoke derisively, had little inclination to seek timely treatment at the hospitals or to notify the government of their sickness. The notification system had failed, as had the short-lived tuberculosis dispensary, although not for good.

Partly due to the lack of popular response, but mostly because of financial reasons, the medical services declared that 'Hospitalisation will not meet the problem of Singapore city'.[63] This was despite growing calls in the newspapers in the late 1930s for a genuine 'tuberculosis service', which would consist of building hospitals for chronic, advanced and incurable cases; sanatoria for early cases; and outdoor clinics for discharged patients. The British deflected these calls, stressing the need to conduct a tuberculosis survey first, or other measures such as tackling urban congestion or improving nutrition.[64] The Municipal Commission concurred that clinics and sanatoria, being beyond the means of the working-class population, would 'barely touch the fringe of our problems'.[65] The official perspective was simply that

> The provision of specialised services, which are often advocated in the press will do little to solve the tuberculosis problem so long as present social standards exist....
>
> In the writer's opinion the remedy is to get rid of over crowding (not necessarily slum clearing) and aim at 'immunity'.[66]

Such views passed the buck back onto the urban building authorities, specifically the Singapore Improvement Trust. But, as the doctors were also aware,

Trust housing was impractical, as their rentals were too high for 'the large mass of people whose income barely keeps them from starvation'.[67] It appeared that the tuberculosis service and urban reform had to go hand in hand. Both likely required substantial political and financial commitment from the state. Before the debate could proceed further, war came upon Malaya, not only disrupting the period of British colonial rule, but also transforming the history of Singapore. It would have major consequences for tuberculosis control.

Notes

1 CO 275/136 1934 Straits Settlements Medical Report, p. 999.
2 CO 273/518 Notes by the Medical Secretary on the Annual Report of the Medical Department, 1921, p. 12.
3 CO 275/124 1929 Straits Settlements Medical Report.
4 Singapore Municipality, *Administration Report 1923*, p. 3.
5 Singapore Municipality, *Administration Report 1925*, p. 9-D.
6 Singapore Municipality, *Administration Report 1926*.
7 *Malaya Tribune*, 14 May 1923.
8 *Malaya Tribune*, 7 November 1923.
9 *Malaya Tribune*, 3 April 1925.
10 *Malaya Tribune*, 3 April 1925.
11 *Straits Times*, 31 August 1925.
12 *Straits Times*, 12 September 1925; 26 September 1925.
13 J. Lornie, Reservation, in Singapore, *Proceedings and Report of the Commission Appointed to Inquire into the Cause of the Present Housing Difficulties in Singapore, and the Steps Which Should be Taken to Remedy Such Difficulties*, Vol. I (Singapore: Government Printing Office, 1918).
14 *Singapore Free Press*, 25 October 1925.
15 CO 273/529 Memo by P.S. Hunter, 3 July 1925.
16 SIT, *Annual Report 1927–1947*, p. 1.
17 CO 275/128 Report titled 'Tuberculosis', 12 October 1931.
18 CO 275/124 1929 Straits Settlements Medical Report; CO 275/131 1932 Straits Settlements Medical Report.
19 Singapore Municipality, *Administration Report 1933*, p. 5-D.
20 CO 275/142 1936 Straits Settlements Medical Report.
21 SIT 70/41 Report of the Committee Appointed to Make Recommendations for the Redevelopment of Certain Crown Lease Land in Singapore.
22 Singapore Municipality, *Administration Report 1939*.
23 SIT 662/31 Memo, 21 July 1931.
24 CO 275/148 1937 Straits Settlements Medical Report, p. 1005.
25 Tuberculosis was frequently misdiagnosed as pneumonia. D.W.G. Faris, 'Some Figures Relating to Tuberculosis in the Straits Settlements', *Journal of the Malayan Branch of the British Medical Association* 1 (3), December 1937, pp. 211–216.
26 D.W.G. Faris, 'Some Figures Relating to Tuberculosis in the Straits Settlements', *Journal of the Malayan Branch of the British Medical Association* 1 (3), December 1937, pp. 211–216.
27 Faris, 'Some Figures Relating to Tuberculosis in the Straits Settlements'.
28 B.M. Jones, 'Summary of Discussion', *Journal of the Malayan Branch of the British Medical Association* 1 (3), December 1937, p. 252.
29 CO 275/114 1925 Straits Settlements Medical Report.
30 CO 275/124 1929 Straits Settlements Medical Report, p. 654.

31 W.G. Huff, 'Entitlements, Destitution and Emigration in the 1930s Singapore Great Depression', *Economic History Review* 54 (3), 2001, pp. 290–323.
32 SIT 70/41 Report of the Committee Appointed to Make Recommendations for the Redevelopment of Certain Crown Lease Land in Singapore.
33 D.M. McSwan, 'The Problem of Tuberculosis with Special Reference to Singapore', *Journal of the Malayan Branch of the British Medical Association* 1 (3), December 1937, pp. 209–211.
34 R.A. Pallister, 'Some Observations on Pulmonary Tuberculosis in Singapore', *Journal of the Malayan Branch of the British Medical Association* 1 (3), December 1937, pp. 231–235.
35 G. Haridas, 'Tuberculosis in Infants and Children', *Journal of the Malayan Branch of the British Medical Association* 1 (3), December 1937, p. 239.
36 W.E. Hutchinson, 'Some Aspects of the Tuberculosis Problem in Singapore', *Journal of the Malayan Branch of the British Medical Association* 1 (3), December 1937, pp. 218–229.
37 David J. Galloway, 'Observations on the Death Rate', *Journal of the Malayan Branch of the British Medical Association* January 1907, p. 3.
38 D.J. Galloway, 'Notes on Tuberculosis', 8 November 1917, in Singapore, *Proceedings and Report of the Commission Appointed to Inquire into the Cause of the Present Housing Difficulties in Singapore*, Vol. I, pp. C8–9; Tenth Meeting, in Singapore, *Proceedings and Report of the Commission Appointed to Inquire into the Cause of the Present Housing Difficulties in Singapore*, Vol. II, pp. B80–82.
39 David J. Galloway, 'Notes on Some Local Aspects of Tuberculosis', *Medical Journal of Malaya* 3 (1), 1928, pp. 2, 3.
40 Michael Worboys, 'Tuberculosis and Race in Britain and its Empire', in Ernst Waltraud and Bernard Harris (eds.), *Race, Science and Medicine, 1700–1960* (London: Routledge, 1999), p. 145.
41 Worboys, 'Tuberculosis and Race in Britain and its Empire'.
42 J. Tertius Clarke, 'Tuberculosis in the Tropics', *Medical Journal of Malaya* 2 (1), 1927, p. 22.
43 D.M. McSwan, 'A Discussion of Tuberculosis in Malaya', *Medical Journal of Malaya* Vol. IV, 1929, p. 126.
44 CO 275/112 1924 Straits Settlements Medical Report, p. 718.
45 CO 275/129 1931 Straits Settlements Medical Report, p. 1012.
46 Lim Boon Keng, 'Tuberculosis Among the Chinese in Singapore', *Journal of the Malayan Branch of the British Medical Association* January 1904, p. 16.
47 Lim Boon Keng, 'Tuberculosis Among the Chinese in Singapore', *Journal of the Malayan Branch of the British Medical Association* January 1904.
48 Lim Boon Keng, Seventeenth Meeting, in Singapore, *Proceedings and Report of the Commission Appointed to Inquire into the Cause of the Present Housing Difficulties in Singapore*, Vol. II.
49 Lim Boon Keng, 'Tuberculosis Among the Chinese in Singapore', *Journal of the Malayan Branch of the British Medical Association* January 1904, p. 19.
50 Lim Boon Keng, 'Tuberculosis Among the Chinese in Singapore', *Journal of the Malayan Branch of the British Medical Association* January 1904, p. 16.
51 Chen Su Lan, 'Opium and Tuberculosis', *Medical Journal of Malaya* Vol. VII, 1932, p. 25.
52 J.W. Winchester, 'Observations on Mortality from Tuberculosis in the Straits Settlements', *Medical Journal of Malaya* Vol. IX, 1934, pp. 182, 186, 187.
53 CO 275/79 1908 Straits Settlements Medical Report, p. 518.
54 CO 275/84 1910 Straits Settlements Medical Report, p. 523.
55 CO 275/148 1937 Straits Settlements Medical Report, p. 1009.
56 CO 275/148 1937 Straits Settlements Medical Report, p. 1029.
57 CO 275/139 1935 Straits Settlements Medical Report, p. 849.
58 CO 275/94 1914 Straits Settlements Medical Report.

59 CO 275/128 Memorandum on the Incidence of Tuberculosis in the Straits Settlements, 31 August 1931.
60 NAS Tan Jiak Kim, Letter from Tan Soo Bin to A.L. Hoops, 24 August 1927.
61 CO 273/655 Memo by A.G.H. Smart, 7 December 1939.
62 CO 275/150 1938 Straits Settlements Medical Report, p. 879.
63 CO 275/150 1938 Straits Settlements Medical Report, p. 856.
64 CO 275/150 1938 Straits Settlements Medical Report, p. 856.
65 Singapore Municipality, *Administration Report 1931*, p. 50-D.
66 CO 275/150 1938 Straits Settlements Medical Report, p. 857.
67 CO 275/150 1938 Straits Settlements Medical Report, p. 856.

4   Struggle for a post-war policy

The Second World War and Japanese Occupation of Singapore were a catalyst in the history of tuberculosis. For one, the period wrought difficult living conditions and shortages of medical care and supplies, increasing the incidence and death rates of the disease. Upon returning to Singapore in 1945, the British administration implemented measures that were in part a response to the deteriorating conditions. But the war had wider ramifications. Its end heralded the advent of medical planning in Singapore and transformed the very nature of imperial rule.

For the first time, the British colonial administration instituted a robust regime of medical programming. The Medical Department urged that '1947 must be the planning year' for a plan that addressed the urgent medical needs of the population.[1] The result was an ambitious ten-year Medical Plan – the first of its kind in Singapore history. The Plan targeted the integration and expansion of the hospital system and focused on the health of children and on acute diseases. Yet the Plan was also controversial: being drafted by Europeans, it came under criticism from Asian physicians and politicians for failing to include their views, and for its scale and cost. The medical programme had to accommodate anti-colonial sentiments precipitated by the war and Singapore's quick fall to a harsh occupying power.

Tuberculosis was initially neglected but became fundamental to the criticisms and subsequent reform of the Medical Plan. As the Medical Department observed in 1948, 'Tuberculosis did not receive anything like the public attention in this part of the world before the war that it does today although the position was serious enough'.[2] The colonial government's doctors were forced to change their stance in a subsequent Tuberculosis Policy memorandum, committing the government to a leading role in the prevention and cure of the disease and the general health of the people. The memorandum also emphasised the health of children and the need to provide relief for sufferers, while resolving to convert Tan Tock Seng Hospital into a sanatorium, an idea firmly opposed before the war.

The war as a catalyst

The unexpected fall of Singapore after a brief battle with Japanese forces ushered in a different experience of colonial rule. The pre-war British regime had been unwilling to commit resources to the control of tuberculosis. The Japanese government which occupied Singapore in February 1942 was likewise vested in its own self-interest, but its rule differed in two ways: in its brevity and in the persistence of wartime conditions which sapped Japan's resources and strength. While intended to be *Syonan* – Japan's 'light of the south' and centre of its Southeast Asian empire – Singapore's medical services deteriorated during the war. So did the health of its people.

Little information exists in the English-language literature on the incidence of tuberculosis during these years, although the 1949 Medical Department report claimed that 'During the Japanese occupation nothing was done for the tuberculosis patient at all'.[3] But the broader context paints a sober picture. The pre-war Medical Department and Sanitary Boards were abolished and replaced by a new agency called Eiseika, while European medical staff were interned and Japanese replaced many of the local staff. Over 100 staff of the Medical Department of all grades lost their lives.[4] The immediate post-war British reports pointed to the gross neglect of preventive medicine during the occupation, with Asian medical staff forced to provide curative treatment as best they could without support from the new regime. Singapore General Hospital was used as a military hospital throughout the Japanese occupation, but TTSH reverted to being a civilian hospital from mid-1943.[5] But the hospitals seemed to have catered to the lower-middle class during this period, not the low-income population.[6]

Paul Kratoska's fine study on the socio-economic history of the Japanese Occupation in Malaya is unfortunately weak on public health. But British colonial reports immediately after the war claimed that the general health of the population deteriorated, with increases in deaths from diseases like malaria and beri beri. A European doctor concluded that 'during the four months July to October 1944 ... the recorded deaths were more than three times as numerous as in the years before the war', and were unlikely to be due to Japanese killings.[7] Kratoska finds that medicines and medical stocks in Singapore were much reduced and what existed was intended for Japanese use.[8] In his oral history interview, Benjamin Chew, a local doctor at SGH, recounted that the Japanese did not provide the medical supplies he and other doctors asked for.[9] The British civil servant Richard John Froude Curtis, interned at Changi Prison, recalled that the Japanese 'could have treated the whole of the Japanese army if they wanted to, but they just didn't give it to us'.[10] People likely turned to folk remedies for their ailments.[11]

Tuberculosis was not mentioned in Kratoska's work, although he notes that the Japanese made vaccination compulsory for other infectious diseases like typhus and smallpox.[12] The post-war British medical reports traced a rise in the number of deaths from tuberculosis during the occupation: from 1,791 in 1941 to 2,172 the following year, which peaked at 3,324 in 1944 before falling to 2,764 in the final year of the occupation.[13] But while this general trend may be correct, the exact figures and source for a period when the British did not rule Singapore must be read with care.

Reformed after the war, the Medical Department initially appeared optimistic about the state of tuberculosis. The 1946 figure of registered deaths for the illness was 1,976 compared to 1,791 five years earlier. This became the official position, that the situation was not worse than in the late 1930s:

> Pulmonary tuberculosis has attracted a lot of attention of late. While the remarkable drop in recorded deaths since the liberation is apparent, it would also appear that the position in regard to this disease is again more or less the pre-war.[14]

But this perspective was quickly found to be amiss and contrary to people's observations. The British themselves could not explain the decline in deaths from other diseases like malaria and beri beri soon after the war ended. They admitted that current statistics were very much estimates, with certifications of deaths continuing to be unreliable, especially in the rural areas.

By 1947, official statistics placed tuberculosis as the leading cause of death in Singapore, with an incidence of 2.35 per 1,000 persons, though this was comparable to the rates of the 1930s.[15] The government hospitals could only admit moribund and early cases, while many others had to be turned away. While still a hospital for acute diseases, TTSH admitted over 200 tuberculosis patients that year, which rose in subsequent years. Although this was only a fraction of the total number of sufferers in the country, it was already about half of the hospital's bed capacity. In 1948, tuberculosis and pneumonia accounted for a quarter of all reported deaths in Singapore, and two-fifths of hospital deaths.[16] Beyond mortality and hospital data, however, little was known about the prevalence of tuberculosis in the community, although the postcolonial government subsequently claimed that 'In the early 1950s, 75% of the population were infected by the time they were 14 years old'.[17]

In addition to the lack of medicine and medical staff during the occupation, the main factors affecting tuberculosis were the aforementioned state of housing and the malnourishment of the population.[18] The latter, caused by shortages of rice and other imported foodstuffs during and immediately after the war, also spurred a rise in the incidence of beri beri. But while the disease was quickly brought under control when the British returned, tuberculosis remained a severe problem.[19] W.J. Vickers, the Director of the Singapore Medical Services, stated:

> Observation would appear to indicate that there was a considerable increase in T.B. over the enemy occupation period which still persists. This is a pointer to the grave under-nourishment which has occurred and which still continues even if in a less obvious form.[20]

Similarly, the British civil servant Humphrey Morrison Burkill recalled that Japanese rule produced 'bad housing, bad food, no medical', a combination which worsened tuberculosis in the town.[21] As Farleigh Arthur Charles Oehlers, a doctor, related,

> You see people walking around with huge ulcers because of malnutrition.... And I think the hospital wards were filled with numbers of tuberculosis

cases. All conditions which were related to lowered resistance brought about by difficult living conditions and insufficient food.[22]

The problem of housing, already severe in the early twentieth century, acquired new dimensions. The brief battle of Singapore did not seriously damage the urban housing stock, but more crucial was the cessation of the work of the Municipal Commission and Singapore Improvement Trust in sanitation and housing. In 1949, the Medical Department echoed the pre-war sentiments: 'The incidence of pulmonary tuberculosis is influenced very considerably by congestion of population in ill-ventilated premises'.[23] The doctor Benjamin Chew recounted a personal visit to Chinatown's shophouses:

> We knew, we went ourselves to see. In one room there would be about ... you know that cubicles right round the four walls of a small room. And they would crowd, go in there and sleep. And all the 30 people in one bedroom, you'll find that maybe half had tuberculosis. And the others would be just catching it.[24]

In 1948, Andrew Morland, a British tuberculosis expert visiting Singapore and Malaya, proffered that the design of shophouses with long, deep rooms and tiny cubicles was ideal for the spread of the disease.[25] He proposed measures that subsequently became policy: the need for compulsory notification of tuberculosis, BCG vaccination, a focus on children, mass radiography, nutritional efforts, fines for spitting, and social security for patients being treated who were unable to sustain their livelihood.[26] On the threat posed by indiscriminate spitting, he said,

> The sputum of the patient with pulmonary tuberculosis is the main source of infection of others. Spitting on the ground or on the floor is a dangerous habit as the sputum dries and blows about in the form of dust to be inhaled or ingested with food by the people. Spitting on the floor of a room used by children is still more criminal as contamination of a crawling child easily occurs.[27]

On housing, the Medical Department concurred: 'In a country which depends on the natural movement of the external air for ventilation and comfort the external openings for the admission of light and air should be down the long sides of the buildings'.[28]

A new housing phenomenon, which first emerged in the 1930s, further grew during and after the Japanese occupation. Unauthorised squatter housing, built at the fringes of the town, was sheltering increasing numbers of people who were unable to rent a space in the shophouses. As the SIT explained, the Japanese colonial government had let out vacant land to housing contractors in order to raise its revenue, while also failing to remove squatters encroaching onto vacant land.[29] The growth of informal settlements, or urban kampongs, at the margins of the town worsened insanitation, overcrowding and spread of tuberculosis through the 1940s and 1950s.

Compounding the housing problems were major demographic changes that occurred at war's end. Families began to bear more children from 1944 at the tail end of the Japanese Occupation. This accelerated, quickly swelling the population of Singapore to over a million people. From 1947 to 1957, the population grew rapidly at a rate of 4.5 per cent per annum, augmented by high birth and fertility rates and the migration of people from Malaya, China and India, many of whom were looking to marry, settle down and form families.

These demographic developments created a large group of infants and young children, to which the official and medical gazes were drawn. In the 1930s, as discussed in the previous chapter, European doctors had pondered the question of tuberculosis among this group should Singapore's population become more settled. Now, the anxieties were apparently being confirmed, feeding into a language of social reform. As the Medical Department noted in 1946, initial surveys of children in schools revealed that 40 per cent were in 'a poor state of nutrition', a vast increase from 5 per cent before the war, and that 1 per cent showed signs of having tuberculosis, previously rare among this group.[30] This number rose to nearly 2 per cent the following year, out of which 3.7 per cent of the girls inspected had tuberculosis, compared to only 0.35 per cent of the boys. The Department surmised that, given that the majority of cases were Chinese, this was owing to the ethnic group favouring the care of male children.[31]

A survey of admissions to the children's ward at SGH from July 1946 to February 1947 found that 13.31 per cent had tuberculosis, a four-fold increase from 3.44 per cent in 1936.[32] The *Straits Times* warned that while tuberculosis death rates in 1948 were not much worse than those before the war, especially taking into account the rapid increase in population, a disturbing new trend was the increase in the disease among school children.[33]

Children became the central social issue of post-war Singapore, traversing into questions of policy for housing, education, family planning, economic development, and health. Singapore's transformation into an industrialised economy was driven by the state's push to create jobs for the young and fast-growing population. The Social Welfare Department, established in 1946 to carry out social services hitherto provided by community organisations, conflated the youth with crime:

> A result of the introduction of Japanese 'co-prosperity' to Singapore was the result of corruption and venality as part of everyday life.... The effect on the youth of the Colony was deplorable. The current dislocation of family life and the closing of schools made matters worse still. Boys and girls became street-loiterers, pickpockets, thieves, burglars, and gangsters.[34]

To government officials, doctors and the educated public, the medical health of children did not merely involve disease, but had wider consequences for crime and social order. *Malaya Tribune* reporters visiting the congested shophouses in the town declared that children would likely grow up in such ostensibly lawless places into 'toughs'.[35] Another visit led the paper to report youngsters 'suffering from [poor] nutrition, scabies, sores, bad teeth, sore eyes, and were often subjected to coughs of various forms', while 'indulging in mock gangster-police fights, brandishing sticks and hurling stones at each other'. The paper highlighted the dire social repercussions, quoting at length a comment by an elderly resident:

> They learn nothing good. They learn now to pick other people's pockets, steal, participate in black market activities, gamble, smoke, spit and curse. This is how they live day by day. They have also lost respect for elders and will resort to anything if the occasion demands.[36]

The dreaded killer

Parallel to the contemporary narratives, oral histories of tuberculosis consistently recall a rampant infectious disease without cure, common among old Chinese people living in Chinatown and signified by their coughing out of bloody sputum.[37] In the oral history interviews, older Singaporeans typically narrated tuberculosis as an integral part of the difficult world they grew up in, spanning the Japanese Occupation and the aftermath – in short as a micro-narrative of post-war Singapore. In Robert Loh Choo Kiat's recollection, the 'expectancy of life is for very short in those days because of things like tuberculosis and other diseases for which we had at the time, they had no cure for'.[38]

Lee Liang Hye, a teenager who contracted tuberculosis, believed that the cause was malnutrition over 'a period of hardship and shortages', beginning with the occupation. When he applied for a job in the civil service after the war, he was given a temporary post instead of a permanent one on account of his illness.[39] In another interview, Toh Peng Koon related how several of his schoolmates died of tuberculosis due to malnutrition during the war; he considered himself lucky to have survived.[40] Some rickshaw pullers who had contracted tuberculosis believed that smoking opium would cure the disease, if only it had the temporary effect of making them feel stronger.[41] But for the most part, tuberculosis was thought to be incurable.

The cultural product of tuberculosis as a prevalent killing disease was the 'wild statements and speculations' circulating among the general population, which came to the notice of the colonial government.[42] The Medical Department found that 'Many wild statements have been made in regard to its incidence in our local population, but the fact remains that beyond crude deaths and death rates no reliable statistics exist in this direction'.[43] There was a growing fear of contracting tuberculosis among the people and consequently, social stigma against the disease. Speaking of the Jaffna community in Singapore, for example, Wilfred Chellapah observed that it was common to avoid social contact or marriage with three social groups that had 'any blemish', namely:

> any members who have married foreigners, who have gone to prisons or who, particularly the diseased, TB, tuberculosis – people were very much afraid of and if there are any members who had TB, these kinds of things that we look into.[44]

In another interview, Vincent Gabriel remembered 'the frightening type of cough, like a whooping of cough' among tuberculosis sufferers. He was fearful of seeing bloody sputum spat onto the ground: 'I was thinking that spittle would dry up and all the germs would fly out and go to everybody (laughs)'.[45]

Interestingly, irrational fears of infection often sat astride a partial understanding of the pathological and environmental roots of tuberculosis. Many people seemed to understand that tuberculosis was an infectious disease caused by a germ and exacerbated by housing conditions, as Brother Joseph McNally remembered:

> it is true that central Singapore was among the most thickly inhabited parts of the earth at that time and tuberculosis was right there. Now, the people of Katong [in the rural east] were more free of tuberculosis than people living in the city.[46]

The overall effect of these mixed perspectives was to intensify the stigma against the disease, rather than for people to rationalise it. This worsened the physical and economic repercussions for the sufferer and their family. As tuberculosis doctor N.C. Sen Gupta explained, the perception of tuberculosis as 'a deadly disease and a very infectious disease' frequently led to the sufferer being 'shunned by other people, by his friends, his compatriots, even his family for the simple reason that he is a source of danger'. This complicated an already difficult economic situation, with many sufferers attempting to hide the illness from their family and employers.[47]

A medical plan sans tuberculosis

It was partly to such narratives of widespread filth, danger and crisis that the late-colonial state responded. When the British returned to Singapore in late 1945, the problems of malnutrition, insanitation and overcrowding immediately precipitated emergency measures. For instance, the Medical and Social Welfare departments organised communal canteens called People's Restaurants to provide milk and soup for people, and free meals in schools for children.[48] But before long an organised and sustained response to the problems was being contemplated. In 1946, the British abolished the Straits Settlements and made Singapore a colony, with Penang and Malacca joining the Malayan Union and subsequently the Federation of Malaya. In addressing the social problems, the colonial establishment instituted surveys and subsequently plans for health, housing, education, social welfare, and economic development.

The response to local conditions was shaped by a metropolitan model of planned development. The colonial administration in Singapore took as its guide and template the 1942 Beveridge Report of wartime Britain, which identified five 'giant evils' of society to be tackled: want, ignorance, idleness, squalor, and disease, the last two of which related to tuberculosis. In the same way the Beveridge Report laid the foundation for Britain's welfare state, the colonial government in

Singapore greatly expanded its socio-economic policies to become a major influence in people's lives.

In public health, the vanguard of state programming was the Medical Plan submitted by W.J. Vickers to the Singapore Advisory Council in February 1947. He emphasised that 'Full planning is required now, and a definite programme should be laid down and completed by the beginning of 1948'.[49] He warned of the dire health of the population: 'Such observation as has been possible points to coming generations of very poor physique, generations far less able to stand the rigour of a normal life, a C.3 population' – 'C.3' being a certification used in Britain of someone as medically unfit for combat duty.[50] The Medical Plan was intended to improve the health of the general population and particularly the poor – no mean undertaking.

Posing a rhetorical question, whether the health service was to remain a 'small organisation dealing only with the minimum of acute disease, or is an adequate hospital service on European lines to be evolved?', Vickers emphatically rejected the former in favour of the latter.[51] Making clear that there would be no return to the government's previous neglect of public health, he asked the government to undertake a major reform of the hospital system and to fund a huge budget of $51 million over five years.

The problem for Singapore was twofold: on the one hand, insufficient beds in the hospitals and a critical shortage of doctors and other staff, and on the other, an increase in hospital admissions. Drawing from earlier critiques of the hospital system by British biochemist Joseph Needham in 1934 and 1939, Vickers declared that 'a complete overhaul is imperative', specifically, 'an expensive and far-reaching hospital reorganisation'.[52] At the end of war, Singapore's hospitals had about 1,000 beds in total, but what was needed by minimum European standards was four times the number.[53] SGH would lead the way in this reform: it would be expanded to 1,000 beds, followed by the building of a second general hospital with similar bed strength. These two projects cost three-fifths of the total budget. Following Needham's recommendation, the hospitals would adopt a unit system, with each having specialised roles: SGH would deal with acute diseases and the Kandang Kerbau Hospital with women's diseases.

Mirroring wider concerns, the Plan was emphatic on the state of infant health and welfare. It sought improvements in outpatient, school and rural services, all of which were deemed to be grossly inadequate. To meet staffing needs, medical education and the training of doctors would also have to be expanded. Medical teaching was being carried out at the King Edward VII College of Medicine, but this would soon be moved to Singapore's first university, the University of Malaya, established in 1949.

But Vickers relegated tuberculosis to a secondary issue. The Medical Plan would focus on acute diseases such as venereal, mental and infectious diseases. He stated, 'Adequate control against epidemic outbreak from without and within cannot be avoided'.[54] He considered tuberculosis to be a chronic disease, for which nothing could be done at the present moment. The illness was a 'vexed question' which had 'to take its proper place in the medical scheme' behind the 'essential medical services' and attack on acute diseases. He assured that the incidence of tuberculosis had returned to pre-war figures. Housing reform to address the overcrowding was difficult and expensive, for 'unless we are prepared to remove sufferers entirely from our slums and organise a vast re-settlement scheme on proper lines, we must move step by step'.[55] Without further funds, it would be sufficient at the moment for a non-government 'Anti-Tuberculosis Society' to run one or two tuberculosis clinics in the town. There would be no sanatorium.

TTSH, the obvious choice for a sanatorium, at the time had 400 beds, half of them used by tuberculosis patients. But while this number would double to 800 over the course of the Medical Plan, the hospital would be integrated into the unit system,

> extended and modernised into a good secondary type of hospital and infirmary, and it should continue to be used for the overflow from the third class wards of the main general hospital – its present chief function.[56]

TTSH would not be converted into a sanatorium. The Plan included the establishment of a sanatorium of 300 beds as 'an additional item for special consideration' but, budgeted for a sum of $5.8 million, this was set aside for a later time.[57]

The Medical Plan soon became controversial. The scale of state intervention, extensive public works and large budget were subjected to much debate and criticism, especially by Asian politicians and doctors. Vickers' call for the people of Singapore to pay for costly health services was unpopular. He was forced to accept a reappraisal of his submission by an expert committee comprised of other European medical officials. In February 1948, a full year after the initial proposal, the committee reaffirmed the $50 million budget but made some significant changes: the Plan was slowed to a period of ten years, with the proposal for a second general hospital scrapped and instead the number of beds at SGH to be increased to 1,500.[58]

This did not satisfy the critics, many of whom viewed the Medical Plan not only through a public health but also a nationalist lens. The war had made many people, including those from the English-educated Asian classes, wary of colonial domination. Self-determination was preferable, indeed vital. Both the British establishment and Asian elite accepted the main tenets of medical reform – the existence of vast social problems, the need for planning, the leading role of the state, and the importance of science and expertise. But they differed on the agents of change – colonial or Asian – and on the salient issues. In early 1948, the Alumni Association of the King Edward VII College of Medicine expressed its unhappiness that the Medical Plan had been solely crafted by Europeans with an apparent emphasis on the well-to-do. It criticised the high costs, while calling for a committee, which would include Asian representatives, to investigate the state of medical planning in Singapore, and for a shift in emphasis to the health of the general population.[59]

In September that year, the already-modified Medical Plan was further scruti-nised by a new Select Committee which, while led by the Financial Secretary, comprised Asian members such as C.C. Tan, a lawyer, and Lim Yew Hock, a trade unionist. Both Tan and Lim were not doctors but professionals and politi-cians; they would become members of the first Singapore Legislative Council in April 1948. In October, Tan urged the government to 'get at the truth which is hidden behind all the façade of excuses put up by the Medical Department'.[60] The Select Committee accepted the main approach and proposals of the Medical Plan but significantly slashed the budget by a third, to $33.5 million.[61]

The tuberculosis policy

Tuberculosis coalesced and enhanced the criticisms of the Medical Policy. It was a disease of the poor, yet Vickers had argued that it had not surpassed pre-war levels, while insisting that the sanatorium could wait. As an example, when Benjamin Chew, a local doctor and Head of one of the two medical units at TTSH, was posted to SGH in 1946, he asked to continue caring for his tuberculosis patients at TTSH. He was summarily told, 'Chew, your cases can go into the streets'; Chew resigned from the government medical service, went into private practice and became one of the founders of the Singapore Anti-Tuberculosis Association, a newly formed non-governmental organisation, the following year.[62]

In November 1947, C.C. Tan made an influential critique of the colonial govern-ment's position in a meeting of the Advisory Council. He reminded the government of its responsibility to improve the health of the people. Warning of severe over-crowding in the town's shophouses, he highlighted that more than a third of the tuberculosis cases in the previous quarter had died from the disease.[63] He called upon the government to build a new tuberculosis hospital. The *Malaya Tribune* followed up Tan's critique with a story of 103 people cramped into a filthy, poorly ventilated three-storey shophouse in the town meant for 30, where 'every one appeared pale and sickly'.[64]

The combination of political and public pressure led the expert committee on the Medical Plan to admit that on tuberculosis, it was 'prepared to bow to really expert opinion from elsewhere'.[65] In June 1948, Vickers was forced to submit a memorandum titled 'Tuberculosis Policy: Singapore' to the Legislative Council. This was a largely defensive paper in which he tried to accommodate the criti-cisms pertaining to tuberculosis while continuing to underline the difficulty of tackling the disease.

The memorandum outlined the order of priorities in combating tuberculosis to be 'housing, adequate nutrition, the establishment of a tuberculosis hospital-sanatorium to serve as a centre for specialist treatment with a first class clinic for outpatients, preventive inoculation, adequate relief schemes, subsidiary sanatoria'.[66] A sound programme would include prevention and cure: the former would con-sist of housing development, child-feeding schemes, inoculation, and circulation of propaganda, while the latter involved increased hospital capacity, including for children, and home-based follow-up treatment for adults.[67]

Despite alluding to a range of measures, Vickers' preferences were clear. Children, relief and housing were top priorities. Malnutrition since the Japanese Occupation had increased the incidence of tuberculosis among children and school pupils. He urged that 'more and more food and a rapidly developing child-feeding department *must* be envisaged as one of the most important and outstanding factors in any present-day anti-tuberculosis scheme in Singapore'.[68] Parents lacking the financial means should receive relief from the state to properly feed and care for their children suffering from tuberculosis. The government would also expand the hospitalisation programme, number of clinics, school medical services, and other health services for children, and have children suspected of tuberculosis undergo X-ray screening.

A substantial part of the memorandum was devoted to the financial matters: the cost of a public housing programme and a relief scheme, like that existing in Britain, to assist patients unable to support their family during the long period of hospitalisation, treatment and rest. The memorandum stressed that housing and relief were interconnected, for without the latter, good housing to which tuberculosis sufferers were relocated would become '"death camps" and slums of another sort which will take other decades to eradicate'.[69] The need for relief is exemplified in the oral history of Lo Hong Ling, who later became a doctor. Lo's father and elder brother, both living under one roof, contracted tuberculosis, likely from Lo's English tutor who had the disease and was coughing incessantly and spitting in their home. There being no effective cure, his father had to return to China to recuperate, before coming back to Singapore when he recovered. This illustrated not only the physical but also economic impact of tuberculosis, for his father was unable to continue with his trade while he was ill.[70]

As tuberculosis was often 'an economic disaster' for the family, the Tuberculosis Policy memorandum outlined a 'full-blooded tuberculosis scheme' to help patients recover on a wholesome diet while unemployed, adapted from an existing scheme in Gibraltar.[71] This concurred with Morland's call for the costs of relief, being too much for voluntary efforts, to be borne by the government.[72] A proposed sum of $240,000, to be allotted to the Social Welfare Department, was deemed requisite for the home treatment of needy patients. This amount would provide for the diet of a patient, estimated at $2 daily, and the rental of one-room or two-room housing ($15–20 and $10–12 per month respectively). The amount of relief would vary depending on whether the patient was single or married. In this relief scheme, which commenced in 1949, applications would be assessed by doctors and almoners, and relief would be provided to all individuals who required it. One thousand cases were estimated to require relief for periods ranging from two months to a year.

On hospitalisation, the memorandum endorsed the recommendation of the expert committee on the Medical Plan which, disagreeing with Vickers, had proposed that TTSH become 'a centre' for tuberculosis treatment and 'eventually should be given over entirely to this disease when a sufficient bed strength has been built up'. The memorandum agreed that TTSH's eventual complement of

800 beds be devoted to the disease over the ten-year period of the Plan. Vickers reminded the government that while the hospital currently only had 356 beds for tuberculosis patients, this was nearly a quarter of the bed strength in Singapore.[73]

Vickers accepted C.C. Tan's proposal for a second sanatorium in addition to TTSH provided it did not 'jeopardise my general scheme'.[74] Conceding he had overestimated the cost of this sanatorium, he submitted a preliminary plan for it to be built on Crown land at Bedok in the east of Singapore. This would push the number of beds for tuberculosis patients to 1,100, just 100 under the number deemed requisite for Singapore's population,[75] although Vickers cautioned that more staff would have to be recruited to care for the number of patients.

But he rejected a proposal for a second tuberculosis clinic by SATA. He argued that the modern diagnostic and treatment clinic being built in TTSH by the Rotary Club, and jointly funded with the government, would suffice. The *Straits Times* was critical of this seemingly inconsistent gesture, likening it to Vickers giving a bouquet to the Club while a raspberry to the association.[76] The Medical Department hailed the Rotary Tuberculosis Clinic, completed in 1949, as 'an up-to-date tuberculosis clinic [built] on the most modern lines' and as 'the first "brick" in the new Medical Plan'.[77]

The memorandum was also less sure about other anti-tuberculosis efforts. Citing the proceedings of the International Tuberculosis Conference in London in 1947 and the visit of tuberculosis expert Dr Parren to Singapore, it cautioned that the vaccine could 'only be used with expert attention on selected groups of the population which are free of the disease'.[78] The Medical Department similarly urged a cautious approach, as 'B.C.G. is a live vaccine and demands very strictest form of control in its manufacture and distribution. In the field it must be handled by experts who know what they are doing'.[79] However, on vaccination, Vickers and the medical doctors in SATA again differed.[80] Newspaper commentaries were doubtful about the universal applicability of BCG and raised questions about its potential harm.[81] Lest it fuel unrealistic public expectations, the memorandum also proposed that the government's anti-tuberculosis propaganda be tampered by what it could actually accomplish.

Subsequently, the Select Committee on the Medical Plan approved the conversion of TTSH to a sanatorium. It doubled the budget for extending the hospital's quarters and facilities, proposed in the Tuberculosis Policy memorandum, from $1.25 million to $2.8 million. Not surprisingly, with C.C. Tan in membership, the Select Committee also supported the building of the second sanatorium with 300 beds. This would be with a much smaller budget as Vickers had conceded, slashed from his original estimation of $5.8 million to $2.2 million.[82] The sanatorium would be equipped with an orthopaedic section. This removed Vickers' original proposal for St. Andrews Orthopaedic Hospital, which treated non-pulmonary tuberculosis and child diseases, to be expanded from 60 to 120 beds.[83]

SGH remained the most important hospital in the final version of the Medical Plan, but the Select Committee had succeeded in defining most, though not all, of the preventive and curative aspects of the anti-tuberculosis policy. In August 1948, as president of the Singapore Progressive Party, C.C. Tan declared that

'The report of the Director of Medical Services has at long last recognised the necessity of a T.B. hospital and sanatorium'.[84] Singapore's pioneer medical plan had largely survived the anti-colonial criticisms, but its final shape had nationalist characteristics.

For the first time in Singapore history, a firm tuberculosis policy had been laid down as part of the Medical Plan, which would guide the control programme. But this had not been without a struggle. The advent of nationalist politics demanded that public health programmes accommodate Asian views and interests. The debates over the scope and specific components of the anti-tuberculosis policy through 1947 and 1948 revealed nationalist concern and critique of colonial bias when it came to the well-being of the general population. In 1949, the Medical Department quipped, 'Tuberculosis is almost a new post-war problem', forgetting its pre-war history of studying and debating the disease.[85]

The colonial government had embraced the preventive and curative aspects of a major anti-tuberculosis programme, with a focus on the health of children, substantial relief for sufferers and the conversion of Tan Tock Seng Hospital into a sanatorium. As Dr Benjamin Chew surmised later in an oral history interview, the immediate post-war period was the moment when the colonial government finally 'woke up' to the need to tackle the disease on a national scale.[86]

Notes

1 Medical Department, *Annual Report 1946*, p. 48.
2 Medical Department, *Annual Report 1948*, p. 3.
3 Medical Department, *Annual Report 1949*, p. 90.
4 Paul H. Kratoska, *The Japanese Occupation of Malaya: A Social and Economic History* (London: C. Hurst, 1998).
5 Medical Department, *Annual Report 1946*.
6 Proceedings of the Legislative Council of Singapore, *The Medical Plan for Singapore*, 18 May 1948.
7 Medical Department, *Annual Report 1946*, pp. 1–2.
8 Kratoska, *The Japanese Occupation of Malaya*.
9 Oral History Centre, National Archives of Singapore, Interview with Benjamin Chew, Reel 3, 20 October 1983, p. 12.
10 Oral History Centre, National Archives of Singapore, Interview with Richard John Froude Curtis, Reel 5, 17 October 1983, p. 4.
11 Kratoska, *The Japanese Occupation of Malaya*.
12 Kratoska, *The Japanese Occupation of Malaya*.
13 Kratoska, *The Japanese Occupation of Malaya*, p. 276.
14 Medical Department, *Annual Report 1946*, p. 21.
15 Medical Department, *Annual Report 1947*, p. 49.
16 Medical Department, *Annual Report 1949*.
17 *Epidemiological News Bulletin*, 11 (1), January 1985, p. 3.
18 Kratoska, *The Japanese Occupation of Malaya*.
19 Oral History Centre, National Archives of Singapore, Interview with Rudy William Mosbergen, Reel 12, 9 June 1994.
20 Proceedings of the Legislative Council of Singapore, *The Medical Plan*, p. C34.
21 Oral History Centre, National Archives of Singapore, Interview with Humphrey Morrison Burkill, Reel 6, 1 October 1999, p. 10.

22 Oral History Centre, National Archives of Singapore, Interview with Farleigh Arthur Charles Oehlers, Reel 4, 1 June 1984, p. 7.
23 Medical Department, *Annual Report 1949*, p. 47.
24 Oral History Centre, Interview with Benjamin Chew, Reel 4, p. 2.
25 Medical Department, *Annual Report 1948*; *Straits Times*, 20 March 1949.
26 *Straits Times*, 17 March 1949.
27 'Report of Dr. A. Morland on Tuberculosis in Malaya', *The Medical Journal of Malaya* 4 (4), June 1950, p. 274.
28 Medical Department, *Annual Report 1950*, p. 55.
29 SIT, *The Work of the Singapore Improvement Trust 1927–1947*.
30 Medical Department, *Annual Report 1946*, p. 26.
31 Medical Department, *Annual Report 1947*.
32 G. Haridas, 'Tuberculosis among Children Admitted to the Children's Ward, Civil General Hospital, Singapore', *Medical Journal of Malaya* 1 (4), June 1947.
33 *Straits Times*, 13 April 1948.
34 Social Welfare Department, *Annual Report 1946*, p. 30.
35 *Malaya Tribune*, 14 October 1947.
36 *Malaya Tribune*, 10 June 1947.
37 Oral History Centre, National Archives of Singapore, Interview with Vincent Gabriel, Reel 6, 10 March 2005.
38 Oral History Centre, National Archives of Singapore, Interview with Robert Loh Choo Kiat, Reel 1, 6 July 2001, p. 7.
39 Oral History Centre, National Archives of Singapore, Interview with Lee Liang Hye, Reel 1, 15 April 1985, p. 6.
40 Oral History Centre, National Archives of Singapore, Interview with Toh Peng Koon, Reel 3, 29 August 1984.
41 Oral History Centre, National Archives of Singapore, Interview with Chen Swee Soo, Reel 4, 25 February 2000.
42 Medical Department, *Annual Report 1948*, p. 78.
43 Medical Department, *Annual Report 1949*, p. 88.
44 Oral History Centre, National Archives of Singapore, Interview with Wilfred Chellapah, Reel 15, 25 November 1983, p. 1.
45 Oral History Centre, National Archives of Singapore, Interview with Vincent Gabriel, Reel 9, 21 April 2005, p. 27.
46 Oral History Centre, National Archives of Singapore, Interview with Joseph McNally, Reel 2, 8 April 1999, p. 1.
47 Oral History Centre, National Archives of Singapore, Interview with N.C. Sen Gupta, Reel 3, 5 February 1999, p. 9.
48 Medical Department, *Annual Report 1946*.
49 Proceedings of the Legislative Council of Singapore, *The Medical Plan*, p. C30.
50 Proceedings of the Legislative Council of Singapore, *The Medical Plan*, p. C29.
51 Proceedings of the Legislative Council of Singapore, *The Medical Plan*, p. C25.
52 Proceedings of the Legislative Council of Singapore, *The Medical Plan*, p. C27.
53 Proceedings of the Legislative Council of Singapore, *The Medical Plan*, p. C29.
54 Proceedings of the Legislative Council of Singapore, *The Medical Plan*, p. C25.
55 Proceedings of the Legislative Council of Singapore, *The Medical Plan*, pp. C33–34.
56 Proceedings of the Legislative Council of Singapore, *The Medical Plan*, p. C32.
57 Proceedings of the Legislative Council of Singapore, *The Medical Plan*, p. C31.
58 Proceedings of the Legislative Council of Singapore, *Modifications to the Plan*, February 1948, p. C45.
59 *Singapore Free Press*, 19 January 1948.
60 *Straits Times*, 20 October 1948.
61 Proceedings of the Legislative Council of Singapore, *Report of a Select Committee of the Legislative Council on the Medical Plan for Singapore*, 19 October 1948.

62 Oral History Centre, National Archives of Singapore, Interview with Benjamin Chew, Reels 6–7, 27 October 1983.

63 *Morning Tribune*, 14 November 1947; *Indian Daily Mail*, 14 November 1947.

64 *Malaya Tribune*, 14 October 1947.

65 Proceedings of the Legislative Council of Singapore, *Report of a Select Committee*, p. C49.

66 Proceedings of the Legislative Council of Singapore, *Tuberculosis Policy: Singapore*, 17 August 1948, p. C212.

67 Proceedings of the Legislative Council of Singapore, *Tuberculosis Policy*, p. C214.

68 Proceedings of the Legislative Council of Singapore, *Tuberculosis Policy*, p. C214.

69 Proceedings of the Legislative Council of Singapore, *Tuberculosis Policy*, p. C214.

70 Oral History Centre, National Archives of Singapore, Interview with Lo Hong Ling, Reel 7, 22 June 2003.

71 Proceedings of the Legislative Council of Singapore, *Tuberculosis Policy*, pp. C216, C217.

72 'Report of Dr. A. Morland on Tuberculosis in Malaya', *The Medical Journal of Malaya*.

73 Proceedings of the Legislative Council of Singapore, *Tuberculosis Policy*, p. C217.

74 Proceedings of the Legislative Council of Singapore, *Tuberculosis Policy*, p. C218.

75 Proceedings of the Legislative Council of Singapore, *Modifications to the Medical Plan*, p. C46.

76 *Straits Times*, 10 September 1948.

77 Medical Department, *Annual Report 1949*, p. 3.

78 Proceedings of the Legislative Council of Singapore, *Tuberculosis Policy*, p. C215.

79 Medical Department, *Annual Report 1947*, p. 56.

80 *Straits Times*, 14 September 1948.

81 *Straits Times*, 8 April 1947.

82 Proceedings of the Legislative Council of Singapore, *Report of a Select Committee*.

83 Proceedings of the Legislative Council of Singapore, *Report of a Select Committee*.

84 *Singapore Free Press*, 31 August 1948.

85 Medical Department, *Annual Report 1949*, p. 12.

86 Oral History Centre, National Archives of Singapore, Interview with Benjamin Chew, Reel 8, 27 October 1983, p. 9.

5   The Tuberculosis Control Unit

The Tuberculosis Policy of 1948 laid out most of the major aspects of tuberculosis control in Singapore after the Second World War. Its implementation as part of the ten-year Medical Plan, the narrative of which unfolds across this and subsequent chapters of the book, was robust, multifaceted and unlike what had been attempted before the war. In 1950, doctors at the Tan Tock Seng Hospital declared that

> There is thus great activity all over the country, which is following the pattern of development of most countries: a determined effort by voluntary agencies, control of policy by Advisory Committees, and a gradual extension of Government Services.[1]

This statement accurately described the history of the anti-tuberculosis programme in the final decade of colonial rule in Singapore.

One important activity was the gradual conversion of TTSH, initially plagued by numerous delays, into a sanatorium with additional beds and staff to provide free treatment for patients with tuberculosis. This development had been resisted by colonial officials and doctors before the Second World War but now became a symbol of British and local commitment to combat the disease among the general population. At the opening of two new blocks of TTSH in 1957, the Minister for Health A.J. Braga linked its new role as a sanatorium to continuing its long tradition as a hospital for diseased paupers.[2] The hospital-based part of the anti-tuberculosis programme charted new ground, assisted by the use of a trio of antibiotic drugs, although it still used the pre-war language of racial immunity and resistance.

The other major achievement in the 1950s was to create the major instruments of tuberculosis control. As in the pre-war period, there was a pressing need for more accurate and comprehensive information on the prevalence of tuberculosis in the community. In 1955, an Australian team of experts, led by Harry Wunderly, conducted an important study of tuberculosis in Singapore. The Wunderly project was historically significant, leading to major policy developments in the late 1950s, particularly the establishment of the Tuberculosis Control Unit (TBCU) as a central agency to maintain a registry of cases and coordinate the control work, the passing of legislation for notification of cases, and Singapore's first case-finding survey for tuberculosis.

The 1948 tuberculosis policy had also highlighted housing reform of the 'cubicle city' as key to the prevention of tuberculosis and for its general benefit to the health of the population. Medical opinion on tuberculosis did not directly influence the building programme of the Singapore Improvement Trust, which greatly expanded in the 1950s. SIT housing was instead driven by a newly emerging set of town and urban planning expertise, but both physicians and urban planners continued to stress to the government the connections between tuberculosis, housing and the urban environment, and the need for full planning. As public housing and public health development conjoined, Singapore was on its way to becoming a planned public housing state in the 1950s.

Towards a centralised programme

Following W.J. Vickers' Tuberculosis Policy memorandum in 1948, a centralised programme slowly took shape. That year, a Government Tuberculosis Advisory Board was formed to implement and coordinate all aspects of the Tuberculosis Policy, comprising representatives from the government and non-governmental bodies, namely the Rotary Club and the Singapore Anti-Tuberculosis Association. By 1953, the Medical Department claimed that 'the Government has developed a 'formidable tuberculosis service'.[3] There were still, however, major obstacles to overcome and important steps to take. They were taken by the colonial administration working together, from 1955 to 1959, with a coalition government led by the Labour Front. The Labour Front government, elected on a limited franchise and renamed the Singapore People's Alliance in 1959, assumed several local portfolios, including health, for which Braga was the minister.

One of the foremost issues was the prevalence of tuberculosis in the population. This was a question of research, information and data. As before, what was better known was the mortality. In the 1950s, mirroring the decade before the war, there was a significant fall in the death rates for pulmonary tuberculosis even as it gained traction as a major social issue and government policy. Using the average figure of 1,714 deaths during 1939 to 1941 as an index of 100, the death rate for the disease dropped dramatically to 57 in 1949, leading the Medical Department to ponder if tuberculosis had become less prevalent after the war. The decline continued year on year into the 1950s, reaching a low of 20 in 1957 – a fifth of the index, which the Department deemed as 'an even more shocking improvement over pre-war rates'.[4] The death rate for pulmonary tuberculosis plunged from 158.2 per 100,000 persons in 1947 to 22.2 in 1959. In numerical terms, the number of deaths fell from 1,976 in 1946 to 1,211 in 1950, and a mere 358 in 1959.

The authorities were quite certain about the general trend of the mortality. Although they knew that some of the deaths were not properly certified, the number of deaths from other causes misclassified as pulmonary tuberculosis clearly outnumbered mistakes the other way round.[5] The declines in gross numbers and the rate were remarkable given the rapid increase of Singapore's population after the war, which rose from 920,000 in 1947 to 1.6 million in 1959. The country's tuberculosis death rates were still considerably higher than those in Britain and the United States in that same period, but there had clearly been a sustained decrease over the long term from the 1920s.[6] More generally, the Medical Department reported in 1952, the trend for tuberculosis paralleled declines in the general death rate and mortality rates for a host of diseases such as 'malaria and unspecified fever, pregnancy and childbirth, premature birth and diseases of infancy'.[7] The department reported that 'The death rate for 1952 is the lowest on record at 11.20 per 1,000 of the population and compares more than favourably with any Western country as a crude death rate'.[8] In 1957, the Ministry of Health (renamed from the Medical Department) attributed Singapore's low death rates to improvements in clinical, public health and sanitary services.

Nevertheless, as was the case two decades earlier, the decline in tuberculosis death rates did not give full assurance to the Medical Department. If anything, it prompted the government and its doctors to read the statistics against the grain to uncover hidden trends and larger issues. In 1949, the department had cautioned that Singapore's death rates were still 2½ times higher than those of Britain, while four-fifths of the deaths were above the age of 30. Although detailed statistics for the pre-war period were lacking, the department remained worried about the apparent increase in tuberculosis among children under 15.[9] In 1955, Dr R.H. Bland, the Director of the Medical Services, warned of tuberculosis that 'its adverse effect on the prosperity of the country lies in its morbidity not in its mortality'.[10] Such concerns persisted into the 1950s and provided the philosophical sanction for the expansion of the anti-tuberculosis programme. In 1957, the Tuberculosis Specialist Dr R.J. Grove-White emphasised that while death rates had fallen, there was no evidence of a decline in new cases, and it was necessary for the government to undertake a 'more purposeful attack' on the disease.[11]

Research into the incidence of tuberculosis among the general population was an initial and necessary part of the control programme. In 1949 to 1950, the government took a rudimentary step with a voluntary survey of its 2,800 civil servants. While it knew that this was a non-representative sample of likely the healthier portion of the population, the survey found 24 persons (0.85 per cent of the sample) with active tuberculosis, or close to an incidence of 1 per cent of the adult population of Singapore with undetected tuberculosis. In addition, only 11 persons (or 0.4 per cent of the sample) tested positive in their sputum. The survey also revealed that accurate information was being blocked by social stigma against tuberculosis, which led to evasion from sufferers. A number of the positive cases had received treatment and been cleared to return to work, but did not make this fact known to the government until they were called up for the survey.[12]

To the public, tuberculosis was a disease in which fact and myth were intertwined. Local doctor Chen Su Lan noted that many tuberculosis sufferers who lived in the kampongs did not consult qualified doctors, but were seen by folk healers and physicians, who typically failed to make a correct diagnosis and instead prescribed herbs or opium for treatment.[13] The state, too, was caught up in this confusion to some extent. In 1950, the Medical Department reported that 'Pulmonary tuberculosis and venereal disease continue to give rise to a good deal of speculation, and forecasts continue to be made of incidence without adequate statistical information'.[14] Five years later, admitting that 'The morbidity rate is still a matter of speculation', the department relied on 'indirect evidence' to estimate an incidence of active pulmonary tuberculosis of 3 per cent of the population.[15] Various other experts provided different estimates ranging from 2 to 8 per cent.[16]

For the government, the crux of the problem of information, or misinformation, on tuberculosis was the absence of a centralised system of case-finding in the community. In 1955, the government invited Sir Harry Wunderly, Director of the Division of Tuberculosis in the Department of Health in Australia, to undertake a study of tuberculosis in Singapore. As Director of Tuberculosis in Australia, Wunderly had launched a comprehensive anti-tuberculosis programme, including a national survey of its incidence and enacting legislation for its notification.[17] Wunderley submitted 'a most helpful report with definite recommendations', which the Medical Department hoped to quickly implement.[18]

Wunderly's study precipitated several important initiatives to centralise the anti-tuberculosis efforts and obtain requisite knowledge of the disease. The first of these was the creation of a new post of Assistant Director of Medical Services (Tuberculosis), similar to Wunderly's role in Australia, in 1957 to coordinate all aspects of the anti-tuberculosis programme. The post elevated tuberculosis to a top position on the state's public health agenda. The first appointee was Dr C.E. Smith, the Medical Superintendent of TTSH.

The following year, an amendment to the Quarantine and Prevention of Disease Ordinance made it obligatory for doctors to notify the Assistant Director of Medical Services (Tuberculosis) of cases of the disease, while also obtaining 'more complete and fuller details of the patient and information in respect of radiologic and laboratory examinations done'.[19] Hitherto, while notification of tuberculosis cases to the municipal officer (subsequently, the city health officer) had been mandatory under the ordinance, it had been difficult to enforce, largely because of the stigma of infection and the economic consequences of notification for the sufferer. In 1949, the authorities had confessed that 'Notification of the disease appears to have been a failure up to date'.[20] The Medical Department in 1955 attributed the problem to

> the diversity of races, difficulty of language, lack of educational and medical facilities in the rural areas, wrong information supplied by the patients or their relatives, non-compliance with notification laws and the acute dearth of medical practitioners.[21]

Alongside the notification, the amendment to the Quarantine and Prevention of Disease Ordinance provided for a 'Central Registry' of cases, to be maintained by the Assistant Director of Medical Services (Tuberculosis) and containing 'all particulars of persons suffering from or who have died from tuberculosis'.[22] The proposed registry was a matter of controversy, underlining the extent to which the state was penetrating into people's lives. John Ede, the chairman of SATA, protested that compulsory notification and registration was coercive for the patient, as it would take place at their homes and thus reveal their illness to the neighbours and community. This may force them underground. The better system, Ede proposed, was the SATA one, where patients voluntarily submitted themselves to be diagnosed at a hospital or clinic.[23] In reply, the Ministry of Health pointed out that tuberculosis notification was not new in Singapore, having been in force between 1918 and 1937, though largely ineffective, and that smaller registries were already being kept by TTSH and the Rotary Singapore Tuberculosis Clinic. Registration, the ministry argued, was also an internationally recognised practice, with the object of persuading sufferers to seek treatment.[24]

In 1958, the government launched the first tuberculosis case-finding programme in Singapore. With assistance from a team of Australian experts supported by the Colombo Plan and led by chest physician Dr Cotter Harvey, a major X-ray screening operation was conducted in four areas, comprising two urban (Chinatown and Jalan Sultan) and two rural (Geylang Serai and Bukit Panjang) areas. The government framed the survey as a fight between the supporters and enemies of Singapore, concerning not only personal health but also the well-being of the whole population:

> We need to defend this young generation from a serious threat – which does not come from outside but from those who live in Singapore and who carry the infection around. We are literally our own worst and most callous enemies.[25]

Publicity material emphasised that X-rays were the 'only way' to detect tuberculosis at an early stage, and that the survey was free, convenient (no undressing was required) and, most crucially, confidential.[26] Those found to have active tuberculosis were urged to seek treatment at TTSH or SATA where, as the government's publicity material stressed, 'with modern drugs doctors can do a lot for you'.[27] Although there were numerous reports that some people would not turn up for the examination as they did not feel they were ill, the response was fairly good, with 50,673 out of 61,622 residents, or 80 per cent, reporting for examination.[28]

The survey discovered that 1,881 persons or 3.7 per cent of the sample had pulmonary tuberculosis. In Chinatown, where 20,000 people were surveyed, 816 active cases of pulmonary tuberculosis were found, or an incidence of 4.08 per cent, while in the less densely populated rural areas, the rate was 1.5 per cent.[29] There were also differences according to age and gender: 3.1 per cent of persons under 50 years of age tested positive, compared to 7 per cent of those over 50, while 5 per cent of males tested positive compared to 2 per cent of females. The survey, while selective, laid the initial basis for the state to estimate the degree of morbidity of tuberculosis in the population.

Finally, Wunderly proposed the formation of a 'central organisation' for 'systematic case-finding in various groups and localities in Singapore', which would help ensure that patients were receiving or continuing with treatment.[30] In 1954, the Medical Department had stressed that the disease would persist until 'there has been developed an organisation efficiently to combat the problem of tuberculosis'.[31] The registry and case-finding work would be organised by the Tuberculosis Control Unit, a self-contained unit which was formed at the end of 1957.

TBCU coordinated all of the state's anti-tuberculosis measures: to formulate plans for controlling tuberculosis, investigate reported cases, monitor patients under treatment and their movement, and follow up with cases at the contact and recall clinics. The unit would also trace contacts, carry out mass X-ray screening, and conduct epidemiological research at its central culture laboratory. It took over the work of the School Tuberculosis Service and the BCG vaccination programme, discussed later in the chapter. Its staff visited patients at their homes, proffering advice on personal and environmental health.[32] The centralised case-finding work more than doubled the number of notifications in Singapore from 2,079 in 1958 to 5,666 in 1959. Most of the notifications were from TTSH, the Rotary Tuberculosis Clinic, and SATA's Royal Singapore Chest Clinic, with a small percentage from private practitioners.

Converting TTSH into a sanatorium

If the government's general tuberculosis policy indicated ambition and robust action, the conversion of TTSH into a sanatorium was one of the more initially halting and disappointing aspects. Still privately run by a Committee of Management, the hospital was intended to be the chief tuberculosis treatment centre as stipulated in the Medical Plan; it was thus no longer subsidiary to the Singapore General Hospital as before the war. The plan provided a larger budget of $2.8 million, increased from the initial $1.25 million, to expand TTSH's capacity to 800 beds. To encourage low-income patients to seek help, tuberculosis treatment would be provided free.

However, the expansion of TTSH turned out to be a protracted and frustrating affair in the first half of the 1950s. The early colonial building efforts focused mostly on the General Hospital, although the overall implementation of the Medical Plan was initially delayed by two years.[33] By the end of 1954, only $5 million out of the plan's projected building budget of $53 million had been spent. Priority in construction had not been given to bed space but to accommodation for the increased numbers of medical and nursing students being trained, who were necessary to care for the heavier volume of admissions. Indeed, severe overcrowding occurred at most of Singapore's hospitals as bed space for general patients shrank.[34]

In particular, the government found it difficult to recruit nursing staff at the minimum ratio of a nurse for every two and a half beds.[35] In the early 1950s, as the Medical Department reported, the growth of education in Singapore and the commercial boom due to the Korean War made nursing a less attractive vocation.[36] In relation to tuberculosis, there was a third factor: the general fear of contracting tuberculosis. As a health attendant who joined TTSH in 1958–1959 admitted, 'At that time, we were afraid we had TB. So we were worried. Every year we have to go for x-ray'. He was able to get himself transferred to the Singapore General Hospital.[37] A proportion of the nurses at TTSH were consequently 'assistant nurses', who were not fully qualified nurses but had adequate training to care for the minority of chronic patients. But the assistant nurses lacked the training to take care of the majority of the hospital's patients, who were treatable cases. TTSH had to recruit its nurses from among the Franciscan Sisters of the Divine Motherhood.

At TTSH, the number of inpatient admissions climbed from 817 in 1948 to 3,362 in 1959, underlining the growing demand for bed space. In 1948, by making use of beds hitherto used for chronic and incurable medical and surgical cases, TTSH was able to increase slightly the number of beds for tuberculosis patients from 326 to 404 – still only half the full amount. Of these beds, 40 were reserved equally for the treatment of boys and girls. In contrast to bed space, of the 817 cases admitted to the hospital in 1948, 352 were discharged as outpatients, 310 cases were treated by collapse therapy and 15 assigned for thoracoplasty. Many of those discharged were turned away due to the lack of beds.

This state of affairs continued into the early 1950s. In 1949, the number of beds for tuberculosis patients was increased by a fifth to 500. This gave Singapore a ratio of 500 beds for tuberculosis patients per million people, still much lower, the Medical Department noted, than Britain's figure of 760. A long waiting list for admission ensued: the hospital was unable to admit 943 cases seen as outpatients due to the lack of beds. Professor E.S. Monteiro, a physician at TTSH, lamented that 'So great however is the number to be treated and so great the reference of cases that the diagnostic side cannot be developed through lack of beds'.[38] In the next few years, 1,000-odd outpatients continued to fail to gain admission every year, and an unspecified number had to be admitted directly from SATA upon the recommendation of the almoner or a medical officer. There were no further expansion works at TTSH until the mid-1950s, with capacity remaining at about 500 beds.

The lack of beds was compounded by the shortage of accommodation for staff. This was deemed to be as equally critical, if not more so, as patient bed space, for without the accommodation, the Medical Department would find it impossible to recruit the additional staff needed. Early building plans for TTSH in 1951 were expressly concerned with quarters and hostels for medical officers, nurses, almoners, health sisters, and hospital assistants.[39] In the early 1950s, although there were 26 wards in TTSH, only 18 were used for tuberculosis inpatients, with three converted into staff accommodation, one into a dental clinic and one into quarters for assistant nurses.

The Medical Plan had provided for a second sanatorium for Singapore with 300 beds. But this failed to materialise. In the Tuberculosis Policy memorandum,

W.J. Vickers had suggested that this sanatorium could be built on Crown land in Bedok. Three years on, however, no action had been taken to do so but the proposed location had changed, as the Medical Department announced:

> Some 500 beds are now provided for tuberculosis alone, as compared with 100 before the war and these will be further increased in 1953 when the 'South Winds' mansion and grounds outside the City so generously donated by Mr. Lee Kong Chian have been turned into the first stage of the 300-bedded sanatorium scheme envisaged under the Medical Plan.[40]

However, the government, after appearing to endorse the South Winds Hotel site, ultimately decided to reject it. The site was at Tanjong Balai in Jurong, some 15 miles from the city centre and costly to travel to, while it would also be difficult to build staff accommodation there.[41] Another site at Clementi Road was also rejected due to difficulties in cost and staffing.[42] In 1955, the government even considered using an unutilised part of the quarantine station on St. John's Island as a sanatorium.[43]

The building programme for the sanatoria began in earnest only in 1953. Instead of two sanatoria, the government and the Committee of Management of TTSH decided to expand the hospital, adding five six-storey ward blocks on 12 acres of vacant land adjacent to the existing premises at Moulmein Road. The new wards would provide 1,200 beds (increased from 1,100 in the Medical Plan), while the 750 beds in the old pavilion-type wards would be retained for the care of chronic tuberculosis cases.[44]

In 1954, six years after the launch of the Medical Plan, construction of the new buildings finally commenced. Two of the high-rise blocks were completed in 1956, heralding the first phase of TTSH's expansion and hailed 'an outstanding example of economy with efficiency'.[45] They added 408 beds, nearly doubling the existing number of beds from 564 to 972. The number of admissions to the hospital jumped from 2,021 to 2,566 in 1957. In 1959, the building programme for TTSH was completed, giving it a bed strength of 1,144. But by this time, the targets of the Medical Plan had become obsolete, with the number of beds for TTSH revised upwards to 1,750. This was still nearly 1,000 beds less than what R.H. Bland deemed to be the minimum requirement of 2,660 beds for tuberculosis patients in Singapore, based on a ratio of 2.5 beds per annual death.[46]

In the 1950s, TTSH was thus understaffed and overworked. In 1950, the number of admissions for tuberculosis nearly doubled over two years to 1,561. For most of the decade, the hospital treated over 2,000 new cases of tuberculosis each year. It only had an establishment of one Tuberculosis Specialist and four medical officers to implement and coordinate the government's tuberculosis policy. Dr R.J. Grove-White, the senior chest physician, was appointed the Tuberculosis Specialist in 1950, a post he held until his retirement eight years later. Already in 1950, TTSH claimed that its work had increased 50 per cent over the previous year and three times over that in 1948; the staff was 'stretched to the limit'.[47]

The use of antibiotics formed the backbone of the treatment of tuberculosis, and especially outpatient treatment, generally following clinical practice in Britain and the United States. The Medical Department was cautious in its approach, particularly on the issue of drug resistance. For the tuberculosis patient, the antibiotics available immediately after the war were streptomycin, sodium para-aminosalicylate (PAS) and isoniazid rimifon. In 1947, the former was still considered an 'experimental drug'.[48] The Medical Department was particularly worried about the use of streptomycin in 'unsuitable' cases and the resistance to the drug, which continued to be tested and used selectively.[49] In 1951, following the 9th Streptomycin Conference in the United States, the department experimented with streptomycin and PAS for suitable outpatients. This proved successful, with only 10 per cent of the treatment subjects developing drug-resistant tubercle bacilli. The following year, streptomycin was used in conjunction with a new drug, isoniazid, in order to minimise resistance.

Streptomycin, PAS and isoniazid were the main anti-tuberculosis drugs in the 1950s, usually used together in some combination and for a longer period (a minimum of one year) than previously. This followed the success of the Edinburgh trials of multi-drug therapy in Britain, pioneered by John Crofton.[50] It owed much to these drugs that the shortage of bed space at TTSH in the early part of the decade did not become a greater problem than it did. The use of these drugs grew in prominence in Singapore in the mid-1950s, overtaking collapse therapy, which fell dramatically in number in the second half of the decade. Inpatient care, rest and recovery in the hospital were deemed to be less effective than the antibiotics, which could be prescribed to outpatients. In any case, rest was feasible only for a small number of inpatients with better means. In 1957, there was an 'appreciable increase' in the expenditure on anti-tuberculosis drugs.[51]

Despite the appearance of the new drugs, the old notion of racial resistance continued to circulate among the Western medical community as before the war. Dr Andrew Morland, the tuberculosis expert who had visited Singapore and Malaya in 1948, placed local ethnic groups onto a global scale of racial resistance according to their urban and housing experience: the Chinese with other highly urbanised groups such as Jews and most Europeans as having the strongest natural resistance, the Malays with the non-urbanised groups like North American Indians and Polynesians with the weakest, and the Indians somewhere in between.[52] SATA also possessed a racial view of tuberculosis:

> Fortunately for us, Singapore – which is in fact a Chinese town – is particularly suitable for domiciliary treatment because of the high resistance of that race to tuberculosis. Every day we see cures which could only be regarded as miraculous in Europe.[53]

A 1954 article on defaulters of outpatient review in the *Medical Journal of Malaya* focused solely on the Chinese, while pondering the differences in the level of resistance to tuberculosis between the indigenous and overseas Chinese.

On its narrow scope, the article reasoned that the inclusion of Malay tuberculosis sufferers would distort the study, as they usually sought treatment only when their illness was fairly advanced.[54] Singapore mirrored other countries in this aspect: although globally the theory of racial immunity and resistance declined in relation to environmental explanations for tuberculosis, it persisted into the 1950s.[55] This brings us to the urban history of post-war Singapore.

Urban planning for the 'cubicle city'[56]

There is a twist to the post-war history of tuberculosis in connection with housing development, which also had a planning dimension similar to the Medical Plan. Overcrowding in the town in the early twentieth century had led to the formation of the SIT, which soon went beyond its original mandate to build a small number of housing for the general population in the 1930s. After the war, housing became a central preoccupation of the British colonial government, and efforts to regulate and sanitise the shophouses and urban kampongs, and rehouse their dwellers in Trust housing, were part of the administration's attack on the 'evil giant' of squalor. Tuberculosis would influence this housing programme, though not in the way we may expect.

In the end, tuberculosis somewhat surprisingly did not play a major or direct role in public housing development. As the Tuberculosis Policy memorandum underlined, in the prevention of tuberculosis, '*bad housing* is known to all to be one of the most important factors – if not the most important – in its spread'.[57] The Medical Department and the doctors continued to associate the disease with poorly ventilated and congested housing.[58] In 1952, the Department expressed surprise that tuberculosis was not more of a threat 'in view of the many overcrowded cubicle dwellings which persist with most primitive kitchen and sanitary arrangements, the many squatter areas, and the thousands of hawkers to be met with in Singapore'.[59]

Such a view showed the urban kampongs to have replaced shophouses as the leading concern of the housing authorities.[60] While a milestone in transforming Singapore from a port city of sojourners to a settled society, the rapid population growth of the post-war years pushed larger families out of the town, where it was no longer possible to increase the capacity of the shophouses, while many new migrants were moving directly into the informal settlements. W.E. Hutchinson, a critic of overcrowding before the war and now promoted to Deputy Municipal Health Officer, was concerned about the health effects of these urban trends, warning 'Whatever may be the killing property of overcrowding [in the Central Area] or the ill-health that may result, it has nothing of the urgency that now exists in the creation of these insanitary kampongs'.[61]

The advice of Hutchinson and other doctors in the colonial service shaped the conclusions of the Singapore Housing Committee, which was formed in 1947 to chart a preliminary plan for the government's response to the housing issue. In underlining the need for good, sanitary housing, the committee made specific reference to tuberculosis. It repeated Hutchinson's pre-war opinion that as a preventive measure, 'proper housing conditions should make extensive sanatoria superfluous'.[62] Medical practitioner K. Kiramathy Pathy also cautioned the committee that 'Overcrowding in insanitary cubicles contributes to the spread of many diseases especially of infants and of lung troubles'.[63] The chairman of the Committee concurred with Vickers and Hutchinson that, as a means of prevention, good housing was superior to sanatoria:

> we hear often today that Government should spend millions on sanatoria and clinics to cure tuberculosis. I would do nothing to prevent anything which would alleviate the lot of consumptives, but money spent on curing might more be profitably spent on prevention and it is of little use to cure a man and then send him back to live again in the circumstances and conditions under which he contracted the diseases.[64]

The Committee duly urged the government to implement a comprehensive building programme, to be undertaken by the SIT. The Trust was thus entrusted with a role it had carried out in the 1930s, one which had not been its original purpose as a sanitary body. The Committee proposed that the Trust be given full powers to build low-cost housing as part of a larger plan of urban and economic development. A major part of its work was 'decentralisation' of the dense population of the town to planned housing and industrial estates in the outlying areas of Singapore.[65]

The SIT never obtained these formal powers but would still build 21,000 units of relatively expensive public housing for the population between 1947 and 1959 – quite a substantial amount for a sanitary agency operating within a British policy of decolonisation. Tuberculosis patients, often with assistance from the Medical Department, were given priority in applying to SIT housing.[66] Reportedly one or two of the residents were dying of tuberculosis every week in 1949.[67] In 1956, up to a third of Trust apartments were occupied by families with at least one tuberculosis patient, although the high rent prevented many patients from taking up the housing.[68] The Housing and Development Board, formed in 1960, possessed the full powers of a housing agency which the SIT did not. The Board effectively built upon the colonial housing programme, removing the urban kampongs, often by building emergency flats on the sites of devastating fires, and then dispersing the shophouse dwellers to satellite towns.[69]

Driving these dramatic developments, however, was not tuberculosis or medical opinion but urban planning expertise. In the new received wisdom, accepted by many states around the world and international non-governmental organisations, housing should come under a rational regime of urban planning. The quintessential planning document was the master plan, which determined the social and economic development of a town or city in relation to the larger region or nation. Urban and town planners, rather than doctors, became experts on housing. They were able to connect housing and urban programmes to issues of citizenship and democracy, which related to the concerns of many late-colonial and postcolonial governments.[70]

An example of urban planning was the arrival of the United Nations Mission of Experts on Tropical Housing, led by US urban planner Jacob Crane, in Singapore and other Asian countries in 1950 and 1951. The Mission urged the British government to undertake a programme of planned housing development to replace insanitary housing. Following this visit, Singapore followed the model of British town planning, particularly as applied to the remaking of post-war London. In the early 1950s, British town planner George Pepler arrived in the city-state to advise the SIT. He played an important role in the formulation of the first Master Plan of Singapore in 1955, which recommended the removal of slum and squatter housing and the establishment of public housing estates and new towns.[71]

Most doctors in Singapore supported the idea of urban planning. In his submission to the Housing Committee, British doctor P.S. Hunter concurred with Vickers and Hutchinson on the benefits for prevention of tuberculosis, that

> control of T.B. will be assisted by the dispersal of the population, and that any money available will be more effectively spent on Housing than on T.B. clinics and sanatoria. The prevention of this disease will be brought about more effectively by improvement in the housing of the population, and dispersal over a larger area.

In Hunter's view, the solution was simple: 'all houses had to be built according to one or other type plan – no deviation, alliation or addition to be permitted'.[72] In 1951, the Medical Department likewise attributed the housing problem to 'the lack of a central planning authority with sufficient power over all our villages to ensure that development takes place on proper lines'.[73]

In the final analysis, tuberculosis did not become a direct or leading factor in the housing and urban reforms that would transform the landscape of Singapore in the 1960s and 1970s. By highlighting tuberculosis prevention to be as crucial as the cure, the doctors nevertheless reinforced the emerging urban planning programme. Across a long period of history spanning the decades preceding and following the Second World War, they had made a sustained argument for the connection between tuberculosis and urban housing.

Through the implementation of the 1948 Tuberculosis Policy and the work of the Tuberculosis Control Unit, tuberculosis was one of three diseases which moved the colonial and Labour Front government to implement robust and far-reaching health and social policies in the 1950s, the others being leprosy and mental illness.[74] In a memorandum submitted to the Singapore cabinet in 1955, the Director of Medical Services R.H. Bland described tuberculosis as a grave economic as well as social problem for Singapore which necessitated an acceleration of efforts. Inaction on this infectious chronic illness, he warned, would be tremendously costly for the country in future.[75]

The centralisation of the anti-tuberculosis programme under TBCU, the conversion of TTSH into a sanatorium and the expansion of public housing all gave birth to a fledgling welfare state in 1950s Singapore. For the British, politically the decade was a period of decolonisation, with the premise that a locally elected government would take over the reins of power. In order to prepare the island for self-rule and eventual independence, the colonial administration deemed it necessary to establish and fund a strong state in the major branches of social governance, including health, education, housing, and social welfare. The Medical Plan served as the blueprint for the state to provide free and substantial healthcare services to the general population, including tuberculosis sufferers. The Medical Department had declared in 1950,

> The Singapore Medical Plan seeks to double the present clinical facilities in all fields and will take us quite a long way on that road to a free service for all so ardently desired by so many of the community.[76]

The 1959 general elections heralded a new phase of the tuberculosis control policy. This was the year the people of Singapore went to the polls to elect a new government for a self-governing state. The People's Action Party, an opposition party without formal links to the colonial government, scored a decisive victory over its main rival, the incumbent Singapore People's Alliance, winning 43 out of 51 seats. Immediately after the elections, the PAP announced a new, invigorated health policy, erasing all colonial precedents, but the Ministry of Health emphasised the grave threat still posed by tuberculosis:

> Singapore still has many health needs. Tuberculosis presents the most serious problem: the Report on the Pilot Survey conducted under the Colombo Plan was ready in 1959 and sets the incidence at 3 per cent of the population.[77]

The statement, nationalistic in tone, mentioned the tuberculosis survey but did not otherwise acknowledge the substantial scale and achievements of tuberculosis control in the 1950s. In the run-up to the elections, A.J. Braga, the Minister for Health, had accused the PAP's proposed health programme as being a mirror image of his government's policy.[78] His was also a politically motivated declaration made ahead of the polls. Nevertheless, there was both continuity and change between the late-colonial efforts and what the postcolonial government now set out to do.

Notes

1 'Report of Dr. A. Morland on Tuberculosis in Malaya', *The Medical Journal of Malaya* 4 (4), June 1950, p. 279.
2 DMS 41/47 Speech at the Opening of Two New Ward Blocks at Tan Tock Seng Hospital on 4 January 1957.
3 Medical Department, *Annual Report 1953*, p. 8.
4 Ministry of Health, *Annual Report 1957*, p. 125.
5 Medical Department, *Annual Report 1952*, p. 116.
6 Ministry of Health, *Annual Report 1957*, p. 34.

7 Medical Department, *Annual Report 1952*, p. 4.

8 Medical Department, *Annual Report 1952*, p. 27.

9 Medical Department, *Annual Report 1949*.

10 DMS 41/47 Part VI Vol. III Memo for the Council of Ministers, 'The Medical Plan', 1955.

11 *Straits Times*, 26 March 1957.

12 Medical Department, *Annual Report 1950*, pp. 115, 117.

13 *Straits Times*, 10 July 1950.

14 Medical Department, *Annual Report 1950*, p. 3.

15 Ministry of Health, *Annual Report 1953*, p. 120.

16 CSO TRY 2149/56 Memo DMS to DFS, 18 June 1956.

17 *Straits Times*, 29 May 1949.

18 Ministry of Health, *Annual Report 1955*, pp. 9–10.

19 Ministry of Health, *Annual Report 1958*, p. 17.

20 Medical Department, *Annual Report 1949*, p. 89.

21 Ministry of Health, *Annual Report 1955*, p. 43.

22 Ministry of Health, *Annual Report 1958*, p. 17.

23 *Straits Times*, 15 March 1957.

24 *Straits Times*, 18 March 1957.

25 National Archives of Singapore, text of a Radio Malaya broadcast by Dr C.E. Smith, Ministry of Health, 22 June 1958, www.nas.gov.sg/archivesonline/speeches/record-details/7b3ebe20-bcf7-11e6-b045-0050568939ad.

26 National Archives of Singapore, broadcast by Dr C.E. Smith, Medical Superintendent and chest physician of TTSH, on tuberculosis survey made over Radio Malaya, 5 June 1958, www.nas.gov.sg/archivesonline/speeches/record-details/7615baa6-bcf7-11e6-b045-0050568939ad.

27 DIS 161/58 Publicity pamphlet, 'Fight Tuberculosis'.

28 E. Hanam, 'Heart Disease from a Case-Finding Tuberculosis Survey in Singapore', *Singapore Medical Journal* 2 (1), March 1961, pp. 3–5.

29 SWD 299/50 Memo, SATA Case-Finding Programme, 27 May 1959.

30 Ministry of Health, *Annual Report 1955*, p. 45.

31 Medical Department, *Annual Report 1954*, p. 10.

32 Ministry of Health, *Annual Report 1959*; TBCU, *Brief Report for 1959 (January–October)*.

33 DMS 41/47 Part VI Vol. I Memo, P.W.D. Estimates 1952', 27 March 1951.

34 DMS 41/47 Part VI Vol. III Memo for the Council of Ministers, 'The Medical Plan', 1955.

35 DMS 41/47 Part VI Vol. III Memo for the Council of Ministers, 'The Medical Plan', 1955.

36 Medical Department, *Annual Report 1950*.

37 Oral History Centre, National Archives of Singapore, Interview with Chew Chin Wah, Reel 1, 29 October 1999.

38 'Report of Dr. A. Morland on Tuberculosis in Malaya', p. 279.

39 DMS 41/47 Part VI Vol. I Memo, 'TTSH Extension', 13 March 1951.

40 Medical Department, *Annual Report 1951*, p. 4.

41 DMS 41/47 Vol. VIII Memo from W.J. Vickers to Colonial Secretary, 8 May 1952; MOH HD 41/47 Memo, 'Tuberculosis in Singapore'; *Straits Times*, 10 March 1953.

42 MOH HD 41/47 Memo, 'Tuberculosis in Singapore'.

43 *Singapore Free Press*, 9 June 1955.

44 HD 41/47 Memo Medical Superintendent, TTSH to Director, Medical Services, 24 June 1954.

45 Ministry of Health, *Annual Report 1956*, p. 5.

46 MOH HD 41/47 Memo, 'Tuberculosis in Singapore'.

47 Medical Department, *Annual Report 1950*, p. 119.

48 Medical Department, *Annual Report 1947*, p. 56.

49 Medical Department, *Annual Report 1950*, p. 123.

50 Ministry of Health, *Annual Report 1958*.
51 Ministry of Health, *Annual Report 1957*, p. 182.
52 'Report of Dr. A. Morland on Tuberculosis in Malaya'.
53 SATA, *The Royal Singapore Tuberculosis Clinic of the Singapore Anti-Tuberculosis Association* (Singapore: D. Moore, 1954), p. 9.
54 Colin McDougall, '3-Year Defaulters from a Tuberculosis Out-Patient Clinic', *Medical Journal of Malaya* 9 (2), December 1954.
55 Christian W. McMillen, *Discovering Tuberculosis: A Global History, 1900 to the Present* (New Haven, CT and London: Yale University Press, 2015).
56 Medical Department, *Annual Report 1951*, p. 141.
57 Proceedings of the Legislative Council of Singapore, *Tuberculosis Policy*, p. C214.
58 Medical Department, *Annual Report 1947*, p. 73; Medical Department, *Annual Report 1949*, p. 47.
59 Medical Department, *Annual Report 1952*, p. 2.
60 Kah Seng Loh, *Squatters into Citizens: The 1961 Bukit Ho Swee Fire and the Making of Modern Singapore* (NUS Press and Asian Studies Association of Australia, Southeast Asia Series, 2013).
61 SIT 348/46, Memo by Deputy Municipal Health Officer to Chairman, SIT, 30 October 1946.
62 Singapore, *Report of the Housing Committee* (Singapore: Government Printing House, 1947), p. 1.
63 SIT 475/47 Letter to C.W.A. Sennett, chairman of Singapore Housing Committee, 26 June 1947, p. 1.
64 SIT 475/47 Notes for Discussion on Housing by Commissioner of Lands, 13 June 1947, p. 3.
65 SIT 475/47 Notes for Discussion on Housing by Commissioner of Lands, 13 June 1947.
66 Medical Department, *Annual Report 1952*, p. 123.
67 SIT 1012/49 Memo from SIT to MIT, 29 November 1949.
68 O.B. Leathart, 'The Almoner's Work in Connection with Tuberculosis in Singapore', Pan-Malayan Tuberculosis Conference, *Transactions of the First Pan-Malayan Tuberculosis Conference* 1–4 November 1956 (Singapore: Government Printing Press, 1957).
69 Loh, *Squatters into Citizens*.
70 Kah Seng Loh, 'Emergencities: Experts, Squatters and Crisis in Postwar Southeast Asia', *Asian Journal of Social Science*, Special Focus: Reframing Modern and Contemporary Southeast Asia: Transnational Connections, Comparisons, and Mobilities, 44 (6), 2016, pp. 684–710.
71 Loh, 'Emergencities'.
72 SIT 475/47 Minutes of Meeting of the Singapore Housing Committee, 14 July 1947, pp. 1–2.
73 Medical Department, *Annual Report 1951*, p. 58.
74 Medical Department, *Annual Report 1952*.
75 DMS 41/47 Part VI Vol. III Memo for the Council of Ministers, 'The Medical Plan', 1955.
76 Medical Department, *Annual Report 1950*, p. 5.
77 Ministry of Health, *Annual Report 1959*, p. 1.
78 *Straits Times*, 17 March 1959.

6   The action programme

Disease was one of the five 'ogres' of the 'subservient society' of postcolonial Singapore, along with poverty, ignorance, squalor, and idleness, which the People's Action Party government targeted upon its decisive victory in the 1959 elections.[1] Presiding over the self-governing state of Singapore, the PAP was committed to a vigorous programme of disease control, particularly tuberculosis, within the political framework of democratic socialism. But while PAP leaders framed a narrative of charting new paths following decades of neglect by the colonial administration, a more accurate interpretation would be of the post-1959 health policy, building upon the programme laid out by the Medical Plan. Even the PAP's language of fighting the 'five ogres' had been taken from the Beveridge Report announced in Britain in 1942.

What really defined the PAP's anti-tuberculosis programme was the essence of the party's name: it was an action-oriented programme, more far-reaching in its social reach and impact than the colonial precedent. Slum clearance and urban renewal occurred alongside public health measures. The 1960s brought quick, dramatic medical results, with decisive falls in morbidity and mortality, heralding a turning point in the history of tuberculosis control. This was partly due to the momentum of the measures launched in the 1950s and particularly the establishment of the Tuberculosis Control Unit in 1957, but the PAP's role in robustly implementing and expanding the existing public health and housing programmes was unquestionable.

Also remarkably, after the 1960s when the battle against tuberculosis looked to have been won, the PAP government did not cease or substantially reduce its efforts. Tan Tock Seng Hospital would return from its role as a sanatorium to a general hospital, but TBCU and the Ministry of Health continued to review, innovate and refine other aspects of tuberculosis control. This reflexive approach of the state was exemplified in the evolution of the BCG immunisation programme and a new focus on the surveillance of the elderly population in the 1970s and 1980s. There was thus no 'end of history' in the narrative of tuberculosis in postcolonial Singapore, but continuing reflexivity and reform.

'TB is on the way out'

The results of the anti-tuberculosis activities in the 1960s – characterised by a mix of policy continuity and change – were hailed as a milestone, resulting in the decisive positive outcomes for which the colonial government had worked towards, but to which the PAP government also made major and sustained contributions. One of the final, and most influential, health policy acts of the colonial government had been to establish a coordinating body in the form of TBCU under an Assistant Director of Medical Services (for tuberculosis). Perhaps the most important decision made by the PAP government was to retain this administrative innovation as the basis of tuberculosis control, with TBCU now directly responsible to the Permanent Secretary of the Ministry of Health. As Dr Chew Chin Hin, who joined TTSH in 1957, observed, there was a strong lobby of physicians based at the ministry – almost akin to a 'TTSH mafia' – who emphasised the importance of tuberculosis control to the government.[2]

The Unit moved quickly to expand its work to cover all of Singapore's districts and population. Underpinning the various facets of this work was the expansion of staff, a long-standing issue in the anti-tuberculosis programme. TBCU recruited more nurses and laboratory technicians to screen and immunise an estimated additional 25,000 babies and 20,000 school children every year. Concurrently, its premises along Moulmein Road served as a training centre for nurses on the public health nursing course. The Mandalay Road Hospital located near TTSH, which was part of the hospital and treated women and children with tuberculosis, also carried out tuberculosis training for nurses and auxiliary staff. The trained nurses received the Tuberculosis Nursing Certificate of Singapore, which was recognised by the British Tuberculosis Association.

Other late-colonial precedents were likewise retained. Tuberculosis remained a notifiable infectious disease for which isolation was not considered mandatory and patients could be treated at home.[3] As before, TTSH continued to offer inpatient and domiciliary treatment for tuberculosis on 'medical and social grounds'.[4] Treatable cases received free inpatient treatment for about four months and outpatient treatment thereafter over at least one and a half years. TTSH had a complement of 1,200 to 1,300 beds in the 1960s, close to the figure in the late 1950s. TTSH's occupational and diversional therapy departments continued to provide job training and temporary work respectively for recovering patients as they did prior to 1959.

TTSH thus retained its role as a sanatorium but within a new administrative framework. In 1961, it ceased to be a statutory corporation and came under the government's direct control. This change was, as the Ministry of Health explained, part of the PAP's move to consolidate the administrative system of Singapore:

> In pursuance of Government's policy to integrate the City Council and other statutory organizations into Government in order to achieve a unified administration of all City and Rural health services under one authority, it was decided that the Tan Tock Seng Hospital which exists as a statutory corporation should be transferred to Government.[5]

From a longer perspective, the nationalisation of TTSH brought to a culmination the expansion since 1873 of the government's role at what was originally a paupers' hospital run by the community. Now the hospital's employees became government servants, although the hospital retained its name in recognition of its founder. TTSH was administratively separate from TBCU, but the two institutions worked closely with each other.

As an example, Boon H. Tang was warded in TTSH for five months of treatment for tuberculosis in 1963. He was 19 then, without cough symptoms and had been diagnosed with the disease at an X-ray examination for admission to Nanyang University. Boon was initially saddened to miss his enrolment into university, but found his stay at the hospital 'very nice' and 'quite enjoyable', especially the delectable food which enabled him to put on five pounds. He was treated in a large ward of 20–30 people, most of whom were elderly with just four or five persons younger than him. There seemed to be considerable trepidation about possible infection at the hospital as the nurses wore masks, although his friends and former classmates visited him without seeming to fear contracting tuberculosis. Boon made a full recovery and left Singapore to pursue his studies in Canada.[6]

Quite quickly after 1959, however, TTSH lost its relevance as a sanatorium. Its open wards, windows and doors harkened to an earlier era of tuberculosis treatment centred around rest, nutrition, and access to air and sunlight, occasionally supplemented by surgery or collapse therapy.[7] This regimen was soon surpassed by the continued use of antibiotic treatment, based on the powerful trinity of streptomycin, PAS and isoniazid. In 1961, the *Singapore Free Press* was confident enough to announce, 'TB is on the way out'.[8] By the middle of the 1960s, tuberculosis began to be brought under control – this was at the dawn of postcolonial rule and within 20 years of the launch of the tuberculosis control programme in 1948. The death rate for pulmonary tuberculosis in Singapore, having dropped significantly in the 1950s, plummeted from 37.5 per 100,000 persons in 1960 to 20 in 1967, 15.7 in 1976, and 6.8 in 1987. The number of repeat visits to TTSH fell from 377,866 in 1959 to 129,745 in 1967, when it included other diseases besides tuberculosis. Tuberculosis was no longer the feared killing disease it was regarded in the 1950s. Chemotherapy played an obvious role in the trends and outcomes, as in other countries, but it was not the only factor.

There were also important policy and administrative reasons for the success. One was the widening net of the notification system. With the central registry maintained by TBCU, the number and rate of new cases rose initially to their highest in 1961: 6,299 cases and 373 per 100,000 persons. Significantly, nearly half of the notifications were made by general practitioners, who received $2 for each case and were required to do so under the 1958 Quarantine and Prevention of Disease (Amendment) Ordinance, with the remainder made by the Singapore General Hospital and SATA. By 1968, even as Singapore's population grew towards two million, the number and rate of new tuberculosis cases had fallen to 3,764 and 189 per 100,000 respectively. In 1976, the number of new cases dropped below 3,000 for the first time in Singapore history since the setting up of the tuberculosis registry in 1957, with a rate of new cases of 124 per 100,000. In 1985, new cases numbered under 2,000 for the first time (under 40 per cent of the 1960 figure), with a rate of 76 per 100,000.

Death certificates remained an important source of new tuberculosis cases throughout the 1960s, but their proportion declined from highs of 8–9 per cent in 1959 and 1960 to 2.7 per cent in 1969. This showed the rigour of the notification system, resolving in large part a major problem during the colonial period.[9] The proportion of notifications made by TBCU and TTSH rose from 42.2 per cent in 1959 to 73.9 per cent in 1969, while those by SATA declined by nearly half from 30.4 per cent to 17.7 per cent. Some 20–25 per cent of the notifications to the former had initially been seen by general practitioners who then referred the cases to the government.[10] In the late 1970s, most of the known cases of bacillary pulmonary tuberculosis in the community were found by general practitioners, while asymptomatic and abacillary cases were usually detected by the public health services.[11] General practitioners were increasingly working with the notification system.

Rehousing the population

Housing development, which had commenced in the 1930s before expanding under the British colonial government after the war, played an indirect but equally vital role in tuberculosis control in the 1960s. The PAP administration placed a heavy emphasis on replacing shophouse and urban kampong dwellings with multi-storey state-built housing for nuclear families. The Housing and Development Board (HDB), a housing statutory board formed in 1960 to replace the Singapore Improvement Trust, took on the mantle of rehousing the low-income population of Singapore.

The main obstacle to the rehousing plan was the lack of suitable land in the city for public housing, but this was resolved by the outbreak of the largest fire in Singapore history at Kampong Bukit Ho Swee in 1961. Like the colonial regime, the PAP held kampongs like Bukit Ho Swee to be 'breeding grounds for disease, crime and fire hazards'.[12] The state purchased the fire site for building emergency public housing flats for the 16,000 fire victims. The housing estate built by the HDB over the fire site not only housed the fire victims, but also squatters from nearby urban kampongs and residents of shophouses in the inner city, called the Central Area. The inferno thus triggered a chain reaction programme of kampong clearance and public housing construction.[13] By 1965, the ring of kampongs on the periphery of the city was being progressively replaced by 51,000 rapidly built flats in HDB estates such as Bukit Ho Swee, housing a quarter of Singapore's population.[14]

The new public housing enabled action upon the chief target of the state's urban redevelopment programme: slum clearance in the Central Area, where the majority of Singapore's population still dwelt. As with the urban kampongs, the government held urban renewal to be partly necessary because of the prevalence of disease and specifically, tuberculosis in the shophouses:

The Central Area slums of modern Singapore are the breeding grounds of disease and crime. The incidence of tuberculosis is higher here than anywhere else on the island, as is the incidence of crime and gangsterism.[15]

In particular, the government wanted to rehouse the residents of shophouses which were overcrowded and 'without privacy, light, proper ventilation, sanitation, or any of the elementary amenities of life' – a statement which echoes W.J.R. Simpson's assessment of the Singapore town six decades earlier.[16] This kickstarted the rehousing of families en masse from the Central Area to the HDB housing estates and new towns in the mid-1960s and 1970s. This was aided by the passing of the 1966 Land Acquisition Act, which enabled the government to acquire land for public development, and the removal of rent control in 1969 to encourage the redevelopment of gazetted areas.

Both the Urban Renewal Unit of the HDB and after 1974 the Urban Redevelopment Authority (URA), which was a full statutory board formed to coordinate and implement redevelopment in the Central Area, framed their work in terms of a benign, necessary and organised programme of urban planning and zoned development.[17] As Teh Cheang Wan, the Minister for National Development, recalled in 1984, urban renewal had transformed Singapore from being a 'gigantic urban slum', afflicted by tuberculosis and other rampant diseases, into a 'modern twentieth century city'.[18] But the exodus of people to the new housing was predicated crucially on the construction of emergency public housing on the sites of former kampongs. The move to HDB flats was initially unpopular as being too far from people's workplace or too high-rise, especially for older Singaporeans, but by the late 1960s and 1970s it had become increasingly accepted and desired by the population, who began to purchase the flats.

Thus the sanitary and housing proposals in the trio of reports – Simpson (1907), the Housing Commission (1918) and the Tuberculosis Committee (1923) – were duly implemented by the postcolonial government. As Governor Laurence Guillemard had said in 1923, housing development and anti-tuberculosis work had occurred side by side, reinforcing each other. The PAP had also made high-rise flats a success, although the Municipal Commission had warned that they would become slums. Of the public housing programme, the Ministry of Health acknowledged that 'Perhaps more than any other non-specific measure this has played a key role in the reduction of tuberculosis in Singapore'.[19] The government was as politically committed to rehousing the population as it was to treating tuberculosis patients. Even critics such as the political scientist Robert Gamer accepted that, while aided by colonial precedents, the success of Singapore's public housing programme 'must partially be attributed to the energy of Prime Minister Lee and his colleagues'.[20] The new, sanitary flats played an indeterminate but no less important role in reducing the spread of tuberculosis among the population.

As the disease declined as a whole, official statistics indicated the average tuberculosis patient in the late 1960s and early 1970s to be disproportionately low- to lower-middle-income, Chinese and male. Although there are no full figures for the period, in 1966, professional, technical, administrative, executive, and managerial occupations comprised a mere 2.6 per cent of all notifications, compared to the largest occupational group, the economically inactive, being those too young or old to work, who made up 25.3 per cent. This was followed by homemakers (usually unpaid and female, 16.9 per cent), craftsmen, production workers and labourers (15.6 per cent), sales workers (8.8 per cent), full-time students (8.4 per cent), transport workers (5.2 per cent), service workers (5.5 per cent), and clerical workers (4.5 per cent).[21]

Although the theory of racial immunity and resistance had disappeared, the PAP government continued to utilise the ethnic classifications from the colonial period. The Ministry of Health still listed and tabulated the patients in ethnic terms: in 1967, 84 per cent of tuberculosis cases were Chinese, which was above their composition of 76.2 per cent of Singapore's population in the 1970 national census. By contrast, Malay patients constituted a mere 9.4 per cent of the cases, which was only three-fifths of the Malays' composition of the total population (15 per cent), while Indians and Pakistanis comprised 5.6 per cent of tuberculosis cases, roughly equal to their proportion of the national population. Males were over-represented, forming 72 per cent of the patients. In the next year, the ratios were similar: 81 per cent Chinese, 9.1 per cent Malay and 7.3 per cent Indian, with 70 per cent of the cases being males.[22] Five years later, males made up 70 per cent of the cases.[23] A decade later, the morbidity rate of respiratory tuberculosis among the Chinese was 1.2 times higher than among the Malays and Indians.[24]

With tuberculosis control largely attained, as early as in 1963 three wards in TTSH began to receive a small but subsequently growing number of adult chest patients in order to relieve the congestion at SGH. In 1965, a Thoracic Surgical Unit was established in TTSH, a sign of how 'the hospital is converting from a single speciality hospital to a multiple-purpose and self-contained hospital'.[25] Two years later, the Ministry of Health announced that 'It is envisaged that in the future the Hospital will turn into a General Hospital with special emphasis on chest diseases and tuberculosis'.[26] By 1970, less than a third of TTSH's patients were being treated for tuberculosis. The hospital's tenure as a sanatorium was a brief one, lasting only a generation.

Clinical trials and transnational connections

But not all the developments in the 1960s charted a history of diminution in tuberculosis control. As the illness declined as a major public health problem, research by TBCU's Central Tuberculosis Laboratory expanded. The laboratory carried out bacteriological examinations on tuberculosis cultures and drug susceptibility tests on *Mycobacterium tuberculosis* for government hospitals and outdoor dispensaries. It also conducted susceptibility tests on tuberculosis-positive sputum cultures from SATA's Royal Singapore Chest Clinic. In 1963, the laboratory carried out an epidemiological study of tuberculosis in Singapore, collecting data on tuberculin reaction patterns among the various age groups in the population.[27] The following year, it started a clinical trial on susceptibility to various drugs, comparing patients treated with isoniazid and PAS and those taking isoniazid with thiacetasone (or thiosemicarbasone).

Particularly notable was the laboratory's growing international network and participation in international trials. Tuberculosis specialists at TBCU and TTSH began to present their research at regional conferences and publish in academic journals such as the *American Review, British Medical Journal, Lancet,* and *Medical Journal of Singapore.*[28] In 1963, TBCU was selected by the World Health Organization to conduct research on the virulence of tubercle bacilli in Singapore together with the British Medical Research Council (BMRC). TBCU also replaced the absolute concentration method for testing drug susceptibility with the resistance ratio method used by WHO and BMRC. In the same year, it received a donation of laboratory equipment from the United Nations International Children's Emergency Fund. In 1965, the laboratory building was extended to enable it to carry out a higher volume of bacteriological research.

In 1966, Singapore hosted WHO's First Regional Tuberculosis Training Course in Asia, while TBCU undertook a joint project with TTSH and BMRC to conduct a chemotherapy trial of thiacetazone, with the laboratory conducting the bacteriological research. The trial found the drug to be ineffective and too toxic for use. In 1969, TBCU and WHO jointly commenced further chemotherapy trials to reduce treatment costs particularly for less-developed countries in the Western Pacific region. In 1975, BMRC commenced a study on short-course chemotherapy using rifampicin and other new drugs for pulmonary tuberculosis. This subsequently became the standard treatment regimen for the disease and proved pivotal in reducing the length of treatment from two years to six months and improving the patient treatment completion rate.

Dr Chew Chin Hin, who chaired the committee in the 1970s, helmed what he deemed 'a very gratifying period of research', including the successful clinical trials of short-course anti-tuberculosis chemotherapy. The research continued past the 1980s until 1991, with local physicians presenting their research on tuberculosis at leading international conferences and publishing in international journals.[29] Each year, TBCU also received a number of WHO Fellows visiting Singapore, who were usually impressed with the Unit's work. According to Dr Andrew Chew, the Medical Superintendent of TTSH in the mid-1960s and who organised many of the international collaborative projects, the BMRC and WHO lauded Singapore's successful tuberculosis control programme as 'a beacon to the world' and especially less-developed countries.[30]

Drug-resistant tuberculosis, which began to receive increasing focus internationally after the war, was recognised as a problem in the 1960s. The Ministry of Health, which received a WHO research grant in 1964 to study the phenomenon, acknowledged that 'A growing problem is the increasing incidence of drug resistant cases of pulmonary tuberculosis'.[31] Second-line drugs such as viomycin, kanamycin, cycloserine, pyrazinamide, and ethionamide were available for use, but because they were more costly and needed to be taken in excess of a year, only one-quarter of the patients with drug-resistant tuberculosis were taking them in 1965.[32]

TBCU also conducted research to address the problem of drug-resistant tuberculosis. In 1965, among the chemotherapy trials it jointly conducted with TTSH, BMRC and the US pharmaceutical company Cyanamid International, one sought to determine the utility of ethambutol for treating drug-resistant tuberculosis and its possible use as a substitute for PAS in primary treatment.

A reflexive state in the 1970s

By the early 1970s, much had been achieved in the nationwide control of tuberculosis, but TBCU did not claim a complete triumph. In 1973, reflecting on 15 years of endeavour, it stated in its annual report, 'Much ground has been covered in the Control of Tuberculosis in Singapore since the founding of this Unit in 1958, but much more work remains to be done'.[33] Both the death rate and number of new cases had fallen to about 50 per cent of the 1958 figures, but TBCU discerned three worrying developments. One of these was a new trend: nearly a quarter of the recent notifications were of teenagers and young adults from the ages of 15 to 24. This was partly due to an intensive case-finding programme among adolescents, who received chest X-rays in their final years of primary and secondary school.

The second trend concerned the elderly and males. TBCU noted that following the notable declines of the 1960s, the tuberculosis death rate had stabilised at around 20 per 100,000 in the early 1970s. In particular, the death rate for elderly patients above 65 had fallen least, compared to other age groups. This was cause for concern given the increase in the proportion of Singapore's elderly population above 60 from 3.8 per cent in 1957 to 6.3 per cent in 1973, and also a trend that would continue. There was also an issue of gender: the ratio of deaths for males and females was 4–5 : 1 and that of notifications was 2–3 : 1. TBCU found a high rate of active tuberculosis among male contacts of patients with active tuberculosis above the age of 40 in 1973.[34] Among males, there also was a low mortality rate under the age of 44, a gradual rise up to age 54 and a steep increase thereafter with the highest peak at age 70 and above. While satisfied that the incidence of tuberculosis had fallen among those aged 15 to 54 in the early 1970s, the Unit warned that 'The drop is less satisfactory among the very young and more elderly'.[35]

Finally, the prevalence of tuberculosis in the general population remained unclear, but there were signs that ongoing transmission rates remained high, as suggested by the large number of outpatients who had no prior BCG vaccination but who were found to have long-standing tuberculosis reactors of six years old. This group had no respiratory symptoms and were detected during the compulsory screening of outpatients who had not received an X-ray examination in the past year.[36]

Although these could be taken as signs that the screening system was working as intended, TBCU was nonetheless sufficiently concerned enough to launch a nationwide prevalence survey of pulmonary tuberculosis in 1975, along with other major diseases such as diabetes and the cardiovascular diseases. This survey, assisted by the WHO medical officer for chronic diseases Dr S. Endo, was the second national tuberculosis survey following the pioneering study in 1958. The new survey covered a random sample of 16,000 people aged 15 and above, comprising 1.11 per cent of the estimated population of this age group in Singapore. The study aimed to gather epidemiological data on tuberculosis which would help to 'map out a strategy for the future'.[37]

The survey revealed the problem of 'a large reservoir of infection in persons with undetected active pulmonary tuberculosis in the community'.[38] A percentage of 1.14 of the sample had active pulmonary tuberculosis based on chest X-rays while 0.46 per cent had positive cultures for *M. tuberculosis* from sputum. The estimates of tuberculosis prevalence were 11.4 per 1,000 for active pulmonary tuberculosis and 32.5 per 1,000 for inactive disease. In demographic terms, active disease prevalence was three times higher in males of all ages than females, while it was twice as high among Chinese males as compared to non-Chinese males; there was no such difference among females. In both sexes, the prevalence increased progressively with age, peaking with the group aged 60 years and older. Worryingly, 82.6 per cent of the active cases were not on treatment, a high percentage similar to other countries in Southeast Asia. Nearly three-quarters of the untreated cases were diagnosed with the disease for the first time.

These findings pointed to issues in the case-finding system. The mass X-ray screening had concentrated on the younger segment of the population, with some 85 to 90 per cent under the age of 40. Although the coverage was extensive, it produced a relatively small yield of active pulmonary tuberculosis – 0.5 per 100 general examinations and under 2 per 100 examinations of cases with respiratory symptoms screened at the outpatient dispensaries. The study concluded that there was 'a need to redefine high risk groups for a more discriminative mass active case finding programme and a shift of the priority age group to the elderly section of the population and especially to the elderly Chinese males'.[39] The detection of 'asymptomatics' was also deemed vital, as previously there had been a heavy reliance on the presence of cough symptoms in detection, but only 17.9 per cent of people with active infection and 15.1 per cent of those with inactive disease presented with such symptoms.[40]

Following the survey, the Department of Tuberculosis Control (DTBC, successor to TBCU in 1975) launched a pilot study for a more discriminative X-ray screening system to detect pulmonary tuberculosis among persons above 45 years residing in six areas of known high disease prevalence in 1978. But the yield was lower than expected due to urban redevelopment projects which led to the geographical dispersal of the population.[41] Ironically, this was a way in which urban development had the negative impact of dispersing tuberculosis patients. Nevertheless, DTBC subsequently increased its case-finding work among people 40 years of age and above (especially males), persons with untreated lung lesions, and children with strong tuberculin reactions.[42] This underlined a major shift in the government's tuberculosis control programme from the general population to high-risk groups.

Regardless of the outcomes, TBCU's review of the anti-tuberculosis programme and its national prevalence survey were characteristic of the Singapore government's reflexive-modernist approach to social governance. To the administration, a programme was not regarded as completed or conclusive but needed to continuously be appraised, revised and improved in order to meet new circumstances and future challenges. Nevertheless, despite its findings in 1973–1976, TBCU appears to have missed or underestimated two salient issues. One was the question of drug resistance which, as it reported in 1973, showed an 'improving picture' over the previous five years. TBCU found that the bulk of drug resistance was to streptomycin, followed by insonaid, while resistance to PAS had gradually diminished. It did warn that a relatively high percentage of patients – 2 per cent of all positive cultures that year – showed an initial resistance to all three primary drugs.[43]

The 1975 prevalence survey found that while 82.8 per cent of the positive cultures were fully sensitive to the three primary anti-tuberculosis drugs, 13.8 per cent were resistant to one drug and 3.4 per cent were resistant to more than one drug.[44] In 1977, laboratory tests showed 86.7 per cent of the *M. tuberculosis* isolates cultured from the sputum of patients with pulmonary tuberculosis to be fully sensitive to streptomycin, PAS, isoniazid, and ethambutol, but 8.5 per cent were resistant to one drug and 4.9 per cent resistant to more than one drug.[45] Still, the official view in the early 1980s was that 'Combined therapy prevents emergence of resistant strains and has been uniformly successful in the management of cases'.[46] In 1986, the Ministry of Health noted that 'Resistance to streptomycin has been on the increase in the last three years, rising from a low of 3% in 1982 to 9.5% in 1983, 13.1% in 1984 and 14.5% in 1985'.[47] Rifampicin was finally placed on routine testing for drug sensitivity in 1980, though in the following year, the ministry repeated the stance that 'Generally, primary resistance is not a problem in Singapore', as resistance to streptomycin alone and to isoniazid-streptomycin combined remained unchanged, with a slight increase in resistance to isoniazid alone.[48]

The other understated issue was the entry of increased numbers of short-term workers, from Malaysia since 1970 and from other Asian countries eight years later, to meet the labour shortage in the manufacturing and construction industries. The foreign workers were screened for tuberculosis and venereal diseases. TBCU's 1973 report noted that nearly 7 per cent of new tuberculosis cases were non-Singapore residents, likely Malaysia, Bangladesh, Thailand, Philippines, and Indonesia where the disease was less well-controlled. Though close to statistics for previous years, this was a relatively high figure and ought to have made a stronger impression on TBCU.

In 1979, the government acknowledged that 'There is also an important external source of infection in visitors and immigrant workers from neighbouring countries with high tuberculosis prevalence'.[49] In the 1980s, about 10 per cent of the new tuberculosis cases were foreigners, though the most common communicable diseases reported among them were malaria, enteric fever and acute viral hepatitis.[50] The Minister for Health Dr Richard Hu admitted that eradicating tuberculosis among foreign workers was difficult given the open nature of the port city and economy. In the period 1977 to 1981, 12 per cent of primary one school entrants had a strong tuberculin test reaction, and among them 30 per cent were born overseas in countries where tuberculosis was highly prevalent.[51] The government briefly attempted to restrict and even end the reliance on foreign labour in the 1980s. But by the 1990s, the number of migrant workers in Singapore was on the rise once more, with important implications for tuberculosis control.

In the 1970s and 1980s, the administrative relationship between TBCU and TTSH underwent a number of changes, reflecting a reduced focus on tuberculosis on the whole. In 1975, TBCU was integrated into TTSH to become the Department of Tuberculosis Control. It continued to maintain the central tuberculosis registry, although it lost the laboratory, which became an independent department within TTSH. Six years later, DTBC was transferred from TTSH to the Primary Health Care and Health Education Division of the Ministry of Health. It took over from the hospital the treatment of active pulmonary tuberculosis, supervising the treatment carried out at eight polyclinics and outpatient dispensaries, in addition to its continuing case-finding and contact-tracing work. In 1988, while DTBC continued its tuberculosis control work, a newly formed Epidemiology Department in the Ministry of Health took over the epidemiological surveillance of tuberculosis, alongside sexually transmitted infections, HIV-AIDS and leprosy.

Functioning under a zoned hospital system, TTSH became one of the major regional hospitals of Singapore in the 1970s, serving the population residing in the eastern part of the island. The number of TTSH's beds reserved for tuberculosis patients was slashed by half to 211 in 1977 and was further reduced in subsequent years as the number of new cases continued to tumble. After 1979, TTSH expanded into new areas such as ear, nose and throat services and renal dialysis. Outpatient treatment, discussed in the next chapter, remained the main form of treatment for tuberculosis.

In the late 1980s, as in Western countries, Singapore's public hospitals, including TTSH, underwent a programme of restructuring and corporatisation. The government embarked on a five-year plan to redevelop the ageing hospitals. Corporatisation would give the hospitals greater autonomy and flexibility in recruiting staff and improve their service quality, as the Ministry of Health explained:

> The hospitals were restructured to enable them to be more flexible and autonomous in management and operations. Restructuring also led to the introduction of commercial accounting, greater cost awareness and financial discipline.[52]

In 1999, the redevelopment of TTSH was completed, and the hospital moved to a new, 1,211-bed hospital complex at Jalan Tan Tock Seng. It housed 15 specialist clinics and various pharmacy and health screening facilities, far removed from its origins of a severely underfunded paupers' hospital in the nineteenth century and the sanatorium of the mid-twentieth century.

Disease of the elderly

The number of new tuberculosis cases and deaths continued to fall throughout the 1980s, though small upward spikes occurred every three to five years. In 1985, two new standard drug treatment regimens for tuberculosis treatment appeared. One was a six-month long, fully supervised short-course chemotherapy using rifampicin and pyrazinamide, while another was a partly supervised, nine-month course with rifampicin alone. Both regimens had proven effective in joint studies by DTBC and BMRC.

Tuberculosis remained the most frequently notified communicable disease and among the top ten causes of death in Singapore in the 1980s. There was some mention of the connection between tuberculosis and acquired immunodeficiency syndrome (AIDS), which became apparent at this time. In an international conference in Singapore in 1986, Dr N.C. Sen Gupta highlighted the increasing threat of latent tuberculosis being activated among AIDS patients with a weakened immune system.[53] Between the discovery of the first case of AIDS in the country in 1983 and 1989, only one patient had both diseases.[54] But the chief concern was tuberculosis as a disease of the elderly as was the case in other developed countries, in particular older men. In 1986, 28 per cent of the new notifications were from 40 to 59 years of age while 29 per cent were aged 60 and above. Two-thirds were males, with the proportion of males climbing with age. Between 1960 and 1989, the average percentage annual fall in tuberculosis incidence rates was much lower among those aged 65 and above (4.1 per cent in males and 2.8 per cent in females), compared to those between the ages of 40 and 44 (6.4 per cent in males and 6.0 per cent in females).[55]

Mortality figures for tuberculosis in Singapore, as in other developed countries such as Japan and Hong Kong, were highest among those 60 years of age and above. In 1984, two-thirds of tuberculosis deaths in Singapore came from the group aged 40 and older, and 90 per cent from those 60 years and older. These statistics, however, were likely to be inflated as some of the deaths attributed to tuberculosis were due rather to the late effects of the disease rather than to active tuberculosis.[56] A study in 1987 of 111 tuberculosis deaths certified by hospital doctors and general practitioners found that less than half – 49 cases – were accurately diagnosed; while 12 patients had died due to the effects of tuberculosis (with the disease inactive at the time of death), 41 deaths were unrelated to tuberculosis (although the disease was present in either its active or inactive form), and nine deaths were found to have no evidence of the disease at all. Neither the hospital doctors (despite their access to the patients and case notes) nor general practitioners were reliable sources, certifying accurately only 39 and ten cases respectively. This brought the true mortality rate for tuberculosis that year down from 6.8 per 100,000 to between 1.9 (counting only the active tuberculosis deaths), or 4.9 when including deaths with no or inadequate hospital records.

In demographic terms, the analysis of the deaths due to active tuberculosis showed that 73 per cent were males compared to 27 per cent females, and 67 per cent were Chinese, 23 per cent Malays and 10 per cent Indians. Nearly two-thirds were elderly patients aged 60 years and above.[57] In contrast, tuberculosis had become rare among children below the age of 15. In 1994, the morbidity rate for this young age group was under five per 100,000. Most of these cases had come into contact with a disease carrier, while in older patients, the disease was likely to have been reactivated rather than recently transmitted. In 1989, the male morbidity rate was highest among the Chinese, although for females, the Malays had the highest rate.

The chief concern at this time was the incidence and mortality rate of tuberculosis among the elderly. In 1984, when DTBC unknowingly echoed colonial sentiments made half a century earlier in warning that 'The disease prevalence in Singapore however is still 3½ times greater than in England and Wales', it was referring to the older population.[58] Early tuberculosis in elderly persons was usually not serious enough to make them seek help or for physicians to make an accurate diagnosis. There was a new environmental factor too: although the government's urban development policy had brought the majority of the population out of the shophouses and kampongs into cleaner public housing flats, the heat and humidity in Singapore led many households to install air-conditioning units, and the re-circulation of air within an enclosed area facilitated the transmission of tuberculosis.[59]

In 1985, the government moved to tackle the problem of tuberculosis among the elderly, offering free chest X-ray screening to all Singapore citizens 45 years and older, including patients of private medical clinics. This age group mostly comprised immigrants born in China and India who came to Singapore in the 1930s and missed the vaccination net after the war.[60] The response from both general practitioners and the age group was, however, disappointing.[61] The tuberculosis expert Dr K. Styblo from the International Union Against Tuberculosis, who visited Singapore in 1986, advised that

> In developed countries, *diagnosis* of the constantly decreasing number of new cases of new tuberculosis is, and will remain until the elimination of the disease, the most difficult problem of tuberculosis control.[62]

In a 1991 study of 120 elderly patients aged 65 and above, 72 per cent were born outside Singapore, mostly in China, though most of them had lived in Singapore for over 30 years. The study concluded that 'the pool of infectious cases remains largely in the elderly', but warned that this age group may ignore their conditions and symptoms, instead attributing them to the effects of old age or cigarette smoking, or may not take their medication reliably due to weak memory or poor mental states. It proposed that family members should supervise their taking of medication, as elderly patients may find it difficult to travel to outpatient dispensaries to be so supervised.[63]

Although it built upon the basis laid by the colonial government, the tuberculosis control programme after 1959 was characteristic of the PAP administration's governance of Singapore. The postcolonial health and housing programmes were more vigorously implemented than before, quickly attaining impressive results. Just as crucially, from the 1970s the government and tuberculosis physicians remained reflexive in discerning and responding to new trends and issues, particularly the growing incidence of tuberculosis among elderly people. On the other hand, there was a continuing struggle to come to terms with both old and new challenges in tuberculosis control, such as increased temporary migration to Singapore and drug resistance.

The next two chapters similarly chart the historical impact, and the balance of change and continuity in other aspects of tuberculosis control among the general population across the colonial and postcolonial periods: namely, outpatient treatment, surveillance and case-finding in the community, and the care and immunisation of infants and school children.

Notes

1 Ministry of Culture, *Democratic Socialism in Action, June 1959–April 1963* (Singapore: Ministry of Culture, 1963), n.p.
2 Kah Seng Loh, Interview with Chew Chin Hin, 8 March 2017.
3 In 1976, tuberculosis came under the Infectious Diseases Act, which replaced the Quarantine and Prevention of Disease Ordinance, and remained a notifiable disease.
4 Tuberculosis Control Unit, *Annual Report 1965*, p. 14.
5 Ministry of Health, *Annual Report 1961*, p. 7.
6 Kah Seng Loh, Interview with Boon H. Tang, 13 April 2017.
7 Kah Seng Loh, Interview with Chew Chin Hin, 8 March 2017.
8 *Singapore Free Press*, 28 July 1961.
9 Ministry of Health, *Annual Report 1968*.
10 Tuberculosis Control Unit, *Annual Report 1973*, Vol. II.
11 Goh Kee Tai, *Epidemiological Surveillance of Communicable Diseases in Singapore* (Tokyo: Southeast Asian Medical Information Center, 1983).
12 Teh Cheang Wan, 'Public Housing in Singapore: An Overview', in Stephen H.K. Yeh (ed.), *Public Housing in Singapore: A Multidisciplinary Study* (Singapore: Singapore University Press for Housing and Development Board, 1975), p. 5.
13 Kah Seng Loh, *Squatters into Citizens: The 1961 Bukit Ho Swee Fire and the Making of Modern Singapore* (NUS Press and Asian Studies Association of Australia, Southeast Asia Series, 2013).
14 Housing and Development Board, *50,000 Up: Homes for the People* (Singapore: Housing and Development Board, 1966).
15 Alan F.C. Choe, 'Urban Renewal', in Ooi Jin-Bee and Chiang Hai Ding (eds.), *Modern Singapore* (Singapore: University of Singapore, 1969), p. 163.
16 Choe, 'Urban Renewal', in Yeh, *Public Housing in Singapore*, p. 89.
17 Choe, 'Urban Renewal' in Ooi and Chiang, *Modern Singapore*.
18 Speech by Teh Cheang Wan at the URA's 10th anniversary dinner, 30 March 1984, www.nas.gov.sg/archivesonline/speeches/record-details/73544c98-115d-11e3-83d5-0050568939ad.
19 Ministry of Health, press statement, 'Anti-Tuberculosis Services', 15 March 1975, www.nas.gov.sg/archivesonline/speeches/record-details/7d2538e5-115d-11e3-83d5-0050568939ad.
20 Robert E. Gamer, *The Politics of Urban Development in Singapore* (Ithaca, NY: Cornell University Press, 1972), p. 48.
21 Ministry of Health, *Annual Report 1967*.
22 Ministry of Health, *Annual Report 1967, 1968*.
23 Tuberculosis Control Unit, *Annual Report 1973*, Vol. II.

24 Goh, *Epidemiological Surveillance of Communicable Diseases*.
25 Ministry of Health, *Annual Report 1965*, p. 189.
26 Ministry of Health, *Annual Report 1967*, p. 93.
27 Tuberculosis Control Unit, *Annual Report 1963*.
28 For instance, two papers based on original research were presented at the Medical Congress in Kuala Lumpur, Malaysia: 'Ethambutol in the Re-treatment of Pulmonary Tuberculosis' by S.A. Yeoh and Johan bin Abdullah, and 'The Use of Second-line Anti-Tuberculous Drugs in the Treatment of Pulmonary Tuberculosis' by J.M.J. Supramaniam and Ng Kwok Choy, Ministry of Health, *Annual Report 1967*, p. 94.
29 Oral History Centre, National Archives of Singapore, Interview with Chew Chin Hin, Reel 6, 7 July 1999.
30 Oral History Centre, National Archives of Singapore, Interview with Andrew Chew, Reel 8, 26 June 1995.
31 Ministry of Health, *Annual Report 1964*, p. 181; Singapore government press statement, 'Anti-Tuberculosis Week', 8 January 1963, www.nas.gov.sg/archivesonline/speeches/record-details/78be4c28-115d-11e3-83d5-0050568939ad.
32 Tuberculosis Control Unit, *Annual Report 1965*.
33 Tuberculosis Control Unit, *Annual Report 1973*, Vol. I, p. 25.
34 Tuberculosis Control Unit, *Annual Report 1973*.
35 Tuberculosis Control Unit, *Annual Report 1973*, Vol. II, p. 5.
36 Tuberculosis Control Unit, *Annual Report 1973*, Vol. I.
37 Ministry of Health, *Singapore 1975 Tuberculosis Prevalence Survey* (Singapore: Ministry of Health, 1978), p. 5.
38 *Epidemiological News Bulletin*, 5 (8), August 1979, p. 34.
39 Ministry of Health, *Singapore 1975 Tuberculosis Prevalence Survey*, p. 32.
40 Ministry of Health, *Singapore 1975 Tuberculosis Prevalence Survey*, p. 40.
41 Goh, *Epidemiological Surveillance of Communicable Diseases*.
42 *Epidemiological News Bulletin*, 11 (1), January 1985.
43 Tuberculosis Control Unit, *Annual Report 1973*, Vol. II, p. 11.
44 Ministry of Health, *Singapore 1975 Tuberculosis Prevalence Survey*.
45 *Epidemiological News Bulletin*, 5 (11), November 1979, p. 47.
46 Goh, *Epidemiological Surveillance of Communicable Diseases*, p. 232.
47 *Epidemiological News Bulletin*, 12 (11), November 1986, p. 71.
48 *Epidemiological News Bulletin*, 16 (1), January 1990, p. 3.
49 *Epidemiological News Bulletin*, 5 (8), August 1979, p. 34.
50 *Straits Times*, 7 October 1983.
51 *Epidemiological News Bulletin*, 8 (1), January 1982.
52 Ministry of Health, *Annual Report 1997*, p. 8.
53 *Business Times*, 3 September 1986.
54 *Straits Times*, 8 October 1989.
55 K.K. Tan, A. Cherian and S.K. Teo, 'Tuberculosis in the Elderly', *Singapore Medical Journal* 32, 1991, pp. 423–426.
56 *Epidemiological News Bulletin*, 17 (4), April 1991, p. 22.
57 B.H. Heng, K.K. Tan, K.W. Chan, and T.H. Tan, 'An Evaluation of 1987 Tuberculosis Deaths in Singapore', *Singapore Medical Journal* 31, 1990, pp. 418–421.
58 DTBC, *Annual Report 1984*, p. 1.
59 *Epidemiological News Bulletin*, 11 (1), January 1985.
60 *Straits Times*, 7 November 1986.
61 *Epidemiological News Bulletin*, 12 (11), November 1986.
62 *Epidemiological News Bulletin*, 13 (9), September 1987, p. 55.
63 Tan, Cherian and Teo, 'Tuberculosis in the Elderly', p. 425.

7 Laboratory of citizenship

The Tuberculosis Policy of 1948 and the 'action programme' of the postcolonial government a decade later not only brought forth a vigorous effort against the disease, but also transformed Singapore in various ways. Tuberculosis control extended far beyond the sanatorium at TTSH to the urban periphery and rural areas, reaching the community at large. More than an expansion in numerical or spatial terms, the programme played a crucial role in the making of modern Singapore as a nation-state. In 1958, the Minister of Health A.J. Braga aptly termed the new Institute of Health 'a laboratory of citizenship of public health in politics – for all will work together in the service of health'.[1] Where the earlier colonial focus was on race, the late-colonial and postcolonial tuberculosis control programmes sought to transform the people into citizens, whose own healthcare would coincide with the development and governance of the nation.

As a citizenship programme, the anti-tuberculosis policy laid out the rules and norms for the cure and prevention of the disease within the framework of the nation-state. The curative aspect encompassed outpatient treatment for treatable cases, supported by an allowance scheme for the needy; and the work of outpatient dispensaries for the people living in outlying areas. The prevention of tuberculosis involved visits by almoners and nurses to the homes of patients and their contacts; mass X-ray screening of selected groups of people and the general population; and extensive public education. The citizenship conferring role of tuberculosis control continued with the launch of the 'directly observed treatment short course' (DOTS) in the Singapore Tuberculosis Elimination Programme in 1997. All these measures extended the social reach of government policy and shaped people's attitudes and behaviour towards the disease, though not with complete success.

Treatment in the community

Before the advent of anti-tuberculosis drugs, tuberculosis was often treated in sanatoria, offering a long period of bed rest and exposure to fresh air and sunlight to aid the patient's recovery.[2] In Singapore, this changed after the Second World War even as TTSH was designated as a sanatorium. Part of the reason for this was effective chemotherapy, but just as important was a comprehensive system of outpatient treatment organised by the state. A state-organised system made sense in post-war Singapore where, compared to Britain, there was a lack of family doctors and general practitioners serving the general population.

Warded patients formed only a small fraction of the total number of tuberculosis patients attending TTSH in the 1950s, although this was also initially due to the shortage of beds. In 1948, TTSH admitted 817 inpatients with tuberculosis, compared to 14,277 outpatients, which comprised both new cases and repeated visits. A decade later, in 1959, the hospital had 3,362 inpatients with tuberculosis and 384,826 outpatients, with the latter directed to the Rotary Tuberculosis Clinic located within its grounds. Only a small number of outpatients attended the clinic regularly to receive active therapy, but this group still rose from over 3,500 persons in 1951 to more than 10,000 four years later.

The Rotary Tuberculosis Clinic, opened in 1950, played an important role in the anti-tuberculosis campaign in providing outpatient treatment and X-ray screening. In 1951, with growing numbers of outpatients, the clinic opened regular sessions in the afternoon and assigned each patient to a doctor. In 1954, the clinic declared that it had reached the limits of its operational capacity, and more clinics serving patients residing far away from TTSH were needed. This contradicted W.J. Vickers' assessment in 1948 that a single tuberculosis clinic would suffice for Singapore. To relieve the congestion at the Rotary clinic, another outpatient clinic was opened in Kallang on the afternoons of Monday to Thursday. In 1959, the Rotary clinic saw 2,685 new cases and 337,866 repeat cases and carried out 90,697 X-ray screenings.

After 1959, outpatient treatment continued to be the thrust of the government's anti-tuberculosis work in the community. Increasing control of the disease brought about a dramatic decline in the number of outpatient attendances at TTSH from 384,826 in 1959 to 139,910 in 1968. This fall was all the more remarkable given a growing number of patients who were treated for other illnesses at the hospital. Greater emphasis was also placed on government outpatient dispensaries operating in the outlying areas of Singapore, where an increasing portion of the country's population was residing. In 1974, the Ministry of Health commenced an operational study code-named 'Outpatient Chest Service' to investigate the use of outpatient dispensaries as primary medical centres for tuberculosis case-finding and treatment, which was affirmed the following year. This service became part of the tuberculosis control programme and was extended to 16 outpatient dispensaries throughout Singapore. In 1980, more than four-fifths of tuberculosis patients were treated at TTSH and the outpatient dispensaries, one-tenth at the Singapore Anti-Tuberculosis Association and a very small number by general practitioners.

Outpatient treatment, while effective on the whole, was not without shortcomings, as there was a small group of outpatients who did not continue or complete their treatment, officially termed 'defaulters'. In 1963, SATA reported a default rate of about 4.6 per cent per month, or about 130 persons, of which 70 per cent were persuaded to resume their treatment.[3] In 1965, the Tuberculosis Control Unit started to make home visits to patients who had defaulted their treatment when contact tracing and treatment came under its purview. The following year, it made 255 visits to the homes of defaulters, and in 1973, about 7 per cent of the home visits were to defaulting patients. Defaulting was not uncommon and occurred even within the bounds of TTSH, as a social worker at the hospital in the 1960s attested:

> some of the patients … find it [taking the anti-TB pills] a real discomfort. They used to throw some of the tablets in the drain behind the wards and staff came to know about it because we found that the drains were choked and that's how we discovered they were throwing.[4]

Treatment allowances and home visits

Another part of tuberculosis control beyond the orbit of the hospital was the government relief scheme. This was one of the most emphatic recommendations of the 1948 Tuberculosis Policy paper, seen as fundamental to outpatient treatment. In 1949, the Social Welfare Department launched a domiciliary relief scheme called the Tuberculosis Treatment Allowance Scheme, to be administered by the Rotary clinic. Under the scheme, needy patients with a 'reasonable chance of recovery' would receive a monthly allowance, provided they refrained from employment to recover at home.[5] The allowance would also provide for an adequate diet for the patient and the family.

Women patients were eligible for the allowance scheme if their husbands were unable to support the household, and their allowance included support for domestic help and school fees for children. The assessment of applicants and the amount of allowance was made by a Tuberculosis Treatment Allowance Advisory Committee, which included members of the Silver Jubilee Fund Committee of Management, the Government Tuberculosis Specialist, the senior almoner at TTSH, and a representative from SATA's Royal Singapore Tuberculosis Clinic.[6]

The allowance scheme was run by a newly formed almoner's division in the Medical Department, comprising the almoner (a medical social worker) and several assistants. In administrative terms, the almoner served as a 'Liaison Officer between the Medical Officer and the Social Welfare Department', bridging the medical and social welfare services.[7] Additional almoners were subsequently recruited to deal with the growing volume of work: in 1956, there were four, assisted by support staff who could speak the local vernaculars, especially the varied Chinese dialects. The role of the almoner and support staff was crucial for a disease closely linked to personal, social, cultural, economic, and environmental factors, advice on which was key to the patient's recovery.

The actual reach and impact of tuberculosis relief was modest yet historically significant. In 1950, an average of 305 patients – a very small proportion of the total number of patients – each received monthly allowances of $62.34. More than 350 were certified fit for work that year. This amount was still higher than the allowances paid out by the Public Assistance scheme, another scheme run by the Social Welfare Department (on which some chronic and advanced tuberculosis patients were registered), and the Silver Jubilee Fund. But the tuberculosis allowances were painfully small compared to the monthly household income of $102 used by the Social Welfare Department to mark poverty in the mid-1950s.[8] Being fixed, the allowances did not adjust to increases in the cost of living, such as during the Korean War boom in Singapore in the early 1950s when recipients, while able to subsist on the allowance, could not maintain a full, nutritious diet.[9] One doctor bluntly called the allowances for illness in general 'disgracefully small'.[10]

Still, the relief scheme expanded. An early survey of the first 500 cases covered under the scheme between 1949 and 1952 found that the monthly number of patients supported rose from 160 to 983, though the average payout increased only slightly from $69.77 to $75.47. Both the number of payments and the cost of the scheme jumped more than eightfold in this period. Most recipients received help for less than a year. Nearly three-quarters experienced an improvement in their health, but 31 of 77 patients were deemed to be untreatable and returned to their home countries. Forty-eight patients had their allowances terminated due to poor progress: their prognosis was not positive – most deteriorated in their health while some died. Nearly two-thirds of the patients returned to work, although a quarter either did not do so or were not known to have done so, probably because they failed to find employment. Of those who returned to work, only two-thirds had their former jobs held for them by their employer. The survey nevertheless concluded that on the whole, the allowance scheme helped patients stabilise their condition during treatment.[11]

By 1954, the almoners were dealing with more cases that needed relief, as attendance at the Rotary clinic rose by 23 per cent from the previous year and new cases by 14 per cent. In 1956, while 867 out of 1,021 recipients had benefited sufficiently from the scheme to pass fit for light work, a significant minority of 12 per cent had passed away, deteriorated in their condition, become non-cooperative, or returned to their home country.[12] In terms of social class, the Social Welfare Department approved allowances for rent arrears for male breadwinners with young children and a range of working-class and lower-middle-class occupations, such as hawkers, fishmongers, stall assistants, labourers, clerks, trishaw riders, and self-proprietors.[13] Chinese recipients were over-represented and Malays under-represented in the scheme but roughly equal to the incidence of tuberculosis by ethnic group: in 1955, 82.9 per cent were Chinese and 8.1 per cent Malays, who formed 75.4 per cent and 13.6 per cent respectively of the general population in 1957.[14] The Labour Front government lauded in 1958 that 'Singapore also probably has the honour of possibly being the only country in Asia to have a most progressive and humane scheme for the treatment of tuberculosis sufferers'.[15]

In the postcolonial period, the almoner's department in TBCU continued to interview all newly registered cases, offering aid and advice, and to manage the Tuberculosis Treatment Allowance Scheme. Information on the relief scheme in the 1960s is sketchy, but the number of interviews conducted by the almoners rose slightly from 20,708 in 1959 to 22,788 the following year before falling to 14,391 in 1964. The number of persons on the allowance scheme did not change greatly, ranging from 1,184 to 2,254 during 1960 to 1963, which were surprisingly lower than the 2,349 and 2,484 cases in 1958 and 1959 respectively. This implied a decline that became more apparent in the latter part of the decade. The number of families on the scheme shrank from 559 in 1968 to just 43 in 1976.

The fall was partly due to the efficacy of antibiotic treatment, and also to the government's policy for non-Singapore citizens, who formed a significant but indeterminate proportion of the patients, to return to their home countries. In 1975, the Ministry of Health declared that 'generous financial aid to patients to allow them to undergo treatment and take sick leave if need be', together with public housing, 'was one of the most far-sighted social schemes which helped to control the spread of the disease in the community'.[16] A story in the press in 1976 related how the almoner helped a male shop assistant to stay in treatment, giving him a monthly allowance of up to three-quarters of his pay, help for rent, food and his children's education, and temporary work in the hospital when his employer refused to have him return to his job.[17]

Besides administering the allowance scheme, the almoner division's casework underpinned the expansion of outpatient and domiciliary treatment in the 1950s, through which state medicine penetrated the private lives and homes of the general population. The issues which the almoners encountered and dealt with were immensely varied and complex. The almoner met with the patient on his or her first visit to the Rotary clinic to explain the nature and treatment of the disease, the need to refrain from employment, changes in lifestyle, and to establish the patient's contacts. In 1955, 60 per cent of all new patients at the clinic were seen by the almoner, who also prepared monthly reports recommending the continuation or termination of allowances, and paperwork for patients returning to their home country. The almoner also helped patients who were still infectious to find temporary boarding for their children, and to find employment for those who had recovered.[18] She had to obtain and appraise intimate information on the patient's lives, including economic, financial and cultural matters relating to personal hygiene, nutrition and environmental sanitation.

As outpatients made their way to the clinic in rising numbers, the almoner division's health visitors and staff nurses deployed to the clinic also made an increasing number of home visits to patients and their contacts. This crucial domiciliary social work ranged from interviewing allowance recipients, tracing contacts, examining children, and vaccinating newborn infants. The health visitors' scheme began in 1950 as a small project but expanded in subsequent years. They made home visits at two-monthly intervals until the patient was certified fit for light work, but uncooperative or deteriorating cases were visited more frequently, with further visits made in the following half a year in case of a possible relapse. In 1951, the Rotary Tuberculosis Clinic's visitors made over 10,000 trips to the homes of tuberculosis patients, a fifth of which were in the rural area.

Like the almoners, health visitors proffered extensive advice on treatment, diet and lifestyle, including sleeping arrangements, isolating utensils, disinfecting places, and disposing sputum. They emphasised the importance of rest and more generally the need to maintain a healthy lifestyle and ways to prevent the spread of tuberculosis. Health visitors reported on the presence at home of 'social ills' – unhygienic and other forms of undesirable behaviour.[19] Their role was not only advisory but had practical effects: they made recommendations for patients residing in extremely poor housing to be admitted to hospital or to qualify for Singapore Improvement Trust's public housing. The visitors also referred adult contacts to the contact clinics for X-ray screening and children to the school clinics for tuberculin testing, while infants under 12 months were sent to the Infant Welfare Department to be vaccinated.

Such work was enormously crucial, complex and challenging. TBCU noted that 'The work of a visiting nurse is of a very personal nature. Much of her success is due to her own personality, patience, experience and her willingness to help'.[20] Dr Andrew Morland likewise observed that

> health visitors with knowledge of the language and customs of the inhabitants of the area are essential. The right type of visitor should be able to do much in the home, by giving advice on hygiene and diet, and by gradually building up confidence in the anti-tuberculosis scheme and the value of contact examination.[21]

Their work among the population required the nurses to 'converse with more or less equal facility in three or four or more Chinese languages, Malay, and English'.[22] *Bahasa Melayu* usually sufficed for communicating with the Malay population, but like the Chinese, the Indian population spoke a number of different languages.

Despite their best efforts, housing conditions in the shophouses or urban kampongs, and the underlying socio-economic factors, often impeded the health visitors' advice on hygiene and sanitation. As one visitor explained, the seemingly straightforward matter of locating a home address tucked amid the numerous shophouse cubicles was exceedingly difficult, while 'in the rural areas their homes are scattered about in the country with no names to paths and numbers to the huts', so 'It is often impossible to locate the house'.[23] Many patients also refused to have their contacts approached or examined, presumably because they did not understand the nature of the disease or the need for contact tracing. Others cited their workload or long travelling distance for not bringing their children to the clinic, or saw no reason to do so because their child seemed healthy.

Home visits were an experience and a spectacle for both parties from very different sociocultural worlds. A European doctor who conducted collapse therapy on patients in their shophouse homes recounted:

> I induced many of these lung collapses in the patients' own homes, assisted by the Health Visitor. What an experience! We would climb up several dark flights of stairs with our apparatus, ultimately discover the patient in a grim, dark cubicle, tenanted by a score of other folks, and there on the hard wooden platform-bed, the lung was collapsed with every crack and gap in

walls and doors a mass of staring faces, all curiosity to view this strange Western treatment.[24]

In 1957, faced with a rising workload, the Ministry of Health deemed it impracticable to make home visits to all patients, who would instead be interviewed at the Rotary clinic, except where it was necessary to assess the living conditions or for deteriorating or uncooperative cases.[25] However, the staffing situation at the almoner's division improved the following year, and a sixth almoner was added to the department in 1959. TBCU was able to resume home visits to all patients and contacts, and 12,316 visits were made to patients and 9,979 visits to contacts that year. Emphasis was given to immediate contacts who ate and slept with the patients and to children under five. Five days after the visit, contacts under 15 years of age were instructed to take a tuberculin test at the Contact Clinic at TTSH or the School Tuberculosis Clinic at the Institute of Health; contacts with negative reactions were given the BCG vaccination while those with positive reactions had chest X-rays performed. Older contacts above 15 underwent chest X-rays to determine the presence of active tuberculosis.[26]

In 1964, contact tracing for tuberculosis was reorganised, partly in response to the prevalence of social stigma against the illness. Now, contacts were examined with their family as a unit to ensure that no one in the family was missed. Nevertheless, 559 of the 4,301 families of new cases that year still could not be traced. Between 1967 and 1975, there was a significant fall in the prevalence of active tuberculosis among household contacts from 55 per 1,000 to 17 per 1,000.[27] But as late as 1975, TBCU admitted that many contacts remained reluctant to make their appointments. These required several home visits and reminders to do so, resulting in a large proportion of contacts not being examined in the year of notification and increasing the likelihood of those with active tuberculosis infecting others.[28]

One of the effects of the mass chest X-ray survey in the 1960s, discussed in the following section, was to give a new dimension to home visits. As the survey was nationwide, it allowed visits to be made to the homes of the general population, not only those who had tuberculosis. In 1960 alone, TBCU nurses involved in the mass X-ray exercise visited over 50,000 homes in Farrer Park district to distribute attendance cards to residents and to urge them to go to an X-ray centre. This was a far greater number than the routine home visits made to tuberculosis patients that year (18,399). These visits were made throughout the course of the mass survey as it reached other districts.

Screening the nation

Prevention of tuberculosis through mass radiography had perhaps the biggest effect in reaching the community. In 1949, chest X-ray screening was introduced in Singapore as a new and cost-effective technique in the early detection of tuberculosis. W.J. Vickers had envisaged its utility for finding tuberculosis among children, while Dr Andrew Morland also supported its use in his visit to

Singapore. The adoption of mass X-ray screening also had an Australian influence, having helped to reduce the incidence of the disease in the country.[29] The Australian team of experts under Dr Cotter Harvey which led the 1958 tuberculosis survey in Singapore also helped to train radiologists and pathology technicians while in the city-state.[30] In the 1950s, X-ray screening was conducted not only at the Rotary Tuberculosis Clinic, but also by mobile teams operating in the rural area and by SATA, which had its own mobile units.

More than a form of technology, the X-ray examination became a social ritual for the people of post-war Singapore. It was a marker of good health and, for adult jobseekers, a requisite passage to stable employment. The number of people screened for employment in the 1950s rose as government agencies and large private firms began to install their own X-ray machines. In 1953, the Medical Services carried out a total of 55,349 chest X-ray examinations, with 36,871 (two-thirds) being conducted at TTSH. This was more than double the number of 21,000 five years earlier, and a vast increase from just 6,000 in 1938.

In 1963, as in the colonial period, the Ministry of Health held that 'Case-finding by Mass X-ray examination has been found by experience in Singapore to be the cheapest and most effective method available'.[31] TBCU continued to offer free screening for adults and adolescents 14 years of age and older, while organising compulsory examinations for selected groups deemed to be at risk: namely, medical personnel working on tuberculosis, children and staff in primary schools and crèches, secondary school leavers, government service applicants, and National Servicemen. In 1972, the list of selected groups was expanded to include hawkers and food-handlers in general, foreign workers applying for temporary work permits and reservists who had completed their National Service.

X-ray screening also became more wide-ranging in the 1960s. The number of X-rays done at the Rotary Tuberculosis Clinic dropped from 90,697 in 1959 to 76,815 in 1966 as mobile units from TBCU and SATA began to carry out screening in the dispersed housing areas, including new public housing estates and towns being built in this time. In 1965, a free screening programme was launched at a careers exhibition, where nearly 5,000 visitors were X-rayed, among whom 194 were suspected of having active pulmonary tuberculosis. In 1969, a static X-ray centre was built at 144 Moulmein Road beside the premises of TBCU, partly to screen referrals from the government outpatient dispensaries. TBCU's mobile X-ray unit also began to visit these dispensaries on fixed days of the week, enabling it to operate more effectively in the community.

Conversely, TBCU's Diagnostic Clinic, previously called the Recall Clinic in the colonial years, played an important role in detecting tuberculosis by following up with cases found to have abnormal chest X-rays. This included those persons detected by the mass X-ray surveys, people with scars in their X-rays (who would be periodically screened) and as-yet inactive cases that would also continue to be assessed. Following the overall trends for tuberculosis, the number of new cases discovered at the Diagnostic Clinic fell from 4,788 in 1961 to 2,278 in 1965 and 985 in 1973. The recall rate also dropped from 10 per cent in 1948 to 5.8 per cent in 1967. In 1973, the clinic still contributed one-third of the new notifications of tuberculosis.

By far, the PAP government's seminal contribution in tuberculosis detection and prevention was a mass X-ray screening exercise, launched by TBCU in 1960. This was part of the geographical expansion of the government's public health and sanitary services, which it announced that year:

> With nearly 40 per cent of the population living in the Rural areas and a large number of housing estates and population centres growing up, urban standards are being enforced in such developed areas.[32]

Characterised by an ambitious effort to exceed the scope and reach of the colonial screening programme, the exercise aimed to cover all of Singapore over a number of years, electoral district by electoral district.

Termed a 'community survey', the mass screening was a systematic, top-down and state-organised operation that penetrated all the residential areas. It was a response to the unwillingness of a substantial section of the public to go to the hospital and clinic for an X-ray.[33] The Health Education Section of the Ministry of Health launched a nationwide promotion campaign, using pamphlets, posters and the mass media (radio and the printed press) to publicise the exercise. The X-ray campaign was also brought to the people through the government's mass organisations operating in the electoral districts, namely, the community centres run by the People's Association, a statutory board formed in 1960, and citizens' consultative committees. Both these organisations propagated the government's campaigns to the grassroots, serving as its eyes, ears and mouthpiece.[34] The ministry stated in 1960 that 'The Mass Tuberculosis X-ray campaign conducted during the year owed its success largely to the Health Education Campaign preceding it'.[35]

That year, TBCU announced its 'immediate plan ... to X-Ray the population of the densely populated parts of the city within a radius of ten miles from Clifford Pier'.[36] This referred to the inner ring of urban shophouses – the alleged 'slums' which W.J.R. Simpson had decried half a century ago. Working with SATA, which provided the mobile X-ray units, TBCU surveyed 15,298 out of an eligible population of 24,515 people in 1960 – a relatively low success rate of just above two-thirds, including a small number (201) of non-residents. But the exercise gathered momentum, and by 1963, all of the inner city districts save one had been covered. That year, cases of tuberculosis discovered during the survey accounted for 18 per cent of all notifications. The following year, the remaining district in the inner city, postal district No. 1 and also the site of the pioneering X-ray project in 1958, was re-surveyed. Of the 8,114 residents screened, 656 had suspicious lung images. Postal district No. 19 in the rural area was also covered that year, finding 485 of 10,615 residents with possible lung tuberculosis. These surveys contributed to nearly one-quarter of all notifications that year.

Henceforth, the X-ray drive moved 'relentlessly' to the outlying areas of Singapore island.[37] It reached the rural and urban kampongs and the new towns and public housing estates, which included Nee Soon, Sembawang, Thomson,

Toa Payoh, and Lorong Engku Aman. As two prongs of a singular policy, the screening campaign followed the arc of urban redevelopment in Singapore. Many of the residents of public housing flats had previously dwelt in kampongs and shophouses, which were undergoing extensive programmes of squatter and slum clearance at that time. Tuberculosis started to diminish as a major killing disease by the mid-1960s, but the mass X-ray campaign continued. Having nearly doubled from 50,152 in 1961 to 96,034 in 1966, the total number of persons screened in Singapore rose to a high of 223,742 a decade later before starting to fall. Within a decade, virtually every person aged 15 and above in Singapore had been mass-screened.[38]

Despite the tremendous effort expended, the efficacy of the mass X-ray campaign is difficult to ascertain, at least in a direct sense. In 1971, TBCU reported that the largest proportion of new tuberculosis cases by far comprised those who had volunteered themselves for screening due to chest symptoms. This was followed by outpatients attending the government clinics, while those detected by the mass survey came in third. As TBCU admitted, 'the great majority of the cases notified in 1970 was not discovered by mass x-ray surveys, but by medical examinations, mainly because of chest symptoms'.[39] It appeared that routine screenings, especially those carried out at government outpatient dispensaries, were more effective (and cost-effective) in detection. Just as important, in addition to government efforts, was the growing willingness and agency of people with chest symptoms to be tested.

Similar observations were made about the relative 'yield' of the various screening methods in the next few years. In 1972, TBCU again found the highest yield of cases from the 'own request' group, although this comprised a mere 0.4 per cent of the people X-rayed by the Unit, followed by people screened at outpatient dispensaries. The community survey was briefly halted that year as TBCU intended 'to give greater publicity to the availability of tuberculosis diagnostic facilities in the Unit to all respiratory symptomatic cases'.[40] But this created another problem, as a 1974 study stated, of a 'significant number' of cases among patients attending the clinics who had no respiratory symptoms.[41] In 1981, the Ministry of Health stated that annual X-rays may pose a health risk to the general population, although those over 40 who have never had one were advised to go for a check-up.[42] Four years later, the government assessed the mass X-ray scheme, finding that 'It involves formidable cost in money, manpower and time. Public response is also poor'.[43] That year, the community casefinding survey, together with biennial chest X-rays for school teachers, finally ceased.

X-ray screening, outpatient dispensaries, home visits, and contact tracing played a big part in extending the state's medical services beyond the city into rural Singapore; the number of tuberculosis notifications in the rural area alone doubled from 613 in 1954 to 1,228 in 1959. All these prongs of tuberculosis control had geographical and social dimensions. On the one hand, patients residing in the rural area sought admission into TTSH or attended the outpatient dispensaries in increasing numbers. On the other hand, an assortment of the state's almoners, nurses, health visitors, radiographers, and BCG vaccination teams brought their diverse work directly to patients, contacts, children, and their families.

Stigma and spitting

The technology of mass screening, while mostly effective, also created the social problem of the 'X-rayed unemployable', that is, persons whose traces of tuberculosis in their X-rays were likely to cause them to be rejected by employers, but who were in fact not infectious.[44] The resultant fear of radiography fed into the existing stigma against tuberculosis. As early as 1951, the Government Tuberculosis Advisory Board recommended that former tuberculosis patients who had sufficiently recovered from the disease, but who still had shadows of tuberculous origin in their X-rays, be regarded as employable on the same terms of service as healthy persons. This would, the board explained, help prevent persons with traces of tuberculosis in the lung who were otherwise healthy from becoming unemployable.[45]

Thus the X-ray not only detected traces of tuberculosis in the body but also inserted the disease into the social consciousness of the body politic. This not only hampered the detection of tuberculosis in individuals, but also the crucial work of contact tracing. Despite the increasing numbers of outpatients, as the Medical Department reported in 1952, an unspecified number stopped attending the Rotary clinic for fear that their employer would find out about their illness, and some of them subsequently suffered a relapse.[46] The department also observed that only a small percentage of the contacts of known tuberculosis patients turned up for X-ray, despite being urged by the health visitors to do so.

At the 1956 Pan-Malayan Tuberculosis Conference in Singapore, Dr B.R. Sreenivasan drew attention to the heightened yet irrational fears of pulmonary tuberculosis which had not existed prior to the Japanese Occupation, but which was a product of an exaggerated faith in the science of mass radiography:

> X-Ray reports are often alarming to patients and one sees patients depressed and deprived of sleep through anxiety over an X-Ray report. It is difficult to convince the ordinary patient that the radiologist is handicapped by the fact that he is looking at shadows, that he has not examined the patient and his report is really meant to help the doctor in charge of the case and not to be the last word in diagnosis.[47]

Sreenivasan gave an example of one patient who had coronary thrombosis, but who deemed this condition to be trivial compared to the possibility that the pain in the chest was caused by tuberculosis.

At the same conference, Dr H.M. McGladdery, a senior surgeon at the Singapore General Hospital, echoed Sreenivasan's concerns about the social and economic repercussions for someone who was known to be a tuberculosis sufferer. The problem, McGladdery stressed, was in part due to the widespread use of X-ray screening:

Socially he is in some degree an outcast. In Singapore, tuberculosis is a disease known and dreaded by all. The people know it is infectious and would avoid a known case....

The consumptive is a marked man.... In certain occupations the consumptive is automatically barred from employment the moment the diagnosis is made. Sailors, school teachers and all who work with children are examples of this class.... Unemployment is commonly the main cause of the patient's suffering and its effect is far more important to the understanding of his case than the details of his lesion on the X-ray plate.[48]

The stigma did not diminish in the face of the expansion of the tuberculosis control programme in the 1950s and 1960s. In one instance, a young man arrived angrily at SATA's Royal Tuberculosis Clinic with an X-ray report in hand. He yelled that he had been turned away from a job because the film showed an image of an old healed scar on his lung. He demanded, 'What's wrong with this place? Do the doctors want tea-money before I can get a clean bill of health?' SATA officials were able to calm him down but felt that 'in some distorted corner of his mind he blames SATA for recording the simple truth about his X-ray picture'.[49]

Part of the state's substantial efforts in public health education was targeted at the stigma, although they were not totally successful. The social fear of infection persisted, fuelled partly by failure to understand that tuberculosis was no longer infectious after several weeks of chemotherapy. In 1955, the Health Education section was formed within the Ministry of Health to educate the people and obtain their cooperation in matters of health, sanitation and disease. The focus fell particularly on the rural population, which hitherto had tended to regard the state's health and sanitary officers as 'medical policemen'.[50] Because 40 per cent of men and 75 per cent of women in Singapore were deemed to be illiterate, the section's efforts dwelt on audio-visual material. Films on the nature of tuberculosis and its treatment and prevention through BCG vaccination were shown to patients before they commenced therapy. The disease also featured prominently in the films, photographs, pamphlets, talks, and other publicity material issued by the section, which was circulated in various languages to schools and to the public.

Public education remained a vital part of the anti-tuberculosis programme after 1959. In 1964, the Ministry of Health organised an Anti-Tuberculosis Week based on the World Health Organization theme, 'No Truce for Tuberculosis'. This included a large-scale exhibition at the Victoria Memorial Hall, supported by mass publicity on radio and television. Nearly 2,000 X-rays were taken of the visitors to the exhibition. At the opening of the event, the Minister for Health Yong Nyuk Lin announced that tuberculosis was no longer the top killer in Singapore, having dropped to sixth place. He took pains to recount how, without the benefit of chemotherapy, his mother and a younger sister and brother had died of the disease.[51] Thereafter the emphasis on tuberculosis lessened. In 1976, tuberculosis was one of six infectious diseases highlighted in the ministry's

'Combat Infectious Diseases' campaign, although the chief focus was on the venereal diseases. These campaigns had a wide reach, but one failing was in overstating the symptoms of tuberculosis, particularly persistent coughs, which some patients did not have.[52]

Nevertheless, the fear of infection proved as resilient as the infection itself. In 1963, the Ministry of Health warned of a large number of unnotified cases:

> Annually there are 2,000 to 3,000 Tuberculosis persons who came up spontaneously for treatment because they are ill. Besides there are many more cases who remain at home or at work to spread the disease without treatment.[53]

Two years later, TBCU carried out the first-ever X-ray exercise in Singapore for all government employees. This was, it reported, 'enthusiastically received from the start and the majority of them came forward to be examined'.[54] A small number – 682 out of 17,336 employees (0.39 per cent) – were suspected of lung tuberculosis. This apparent willingness to be screened seemed to suggest a change from the anxiety of becoming an X-ray unemployable in the 1950s, but evidence of people continuing to avoid screening and treatment suggests that some of the fear persisted.

Two decades after the formation of the Health Education section, stigma was still a social concern, as highlighted by a 1974 newspaper article of a young patient who struggled to find permanent employment.[55] Likewise, Dr Goh Kee Tai, Head of the Quarantine and Epidemiology Department of the Ministry of the Environment, observed,

> Many people in Singapore still consider TB a stigma and are reluctant to come forward for a chest X-ray. Therefore, case detection through self selection of the respiratory symptomatics on a voluntary basis has a low yield. Also, a sizeable proportion of the TB population do not appear to have symptoms to feel the need to use the screening facilities widely available in Singapore.[56]

The problem of stigma-induced avoidance, the Ministry of Health noted with concern in 1985, was particularly acute for 'the aged males among whom the disease is known to be high'.[57]

Even though it played an important role in treating tuberculosis, the hospital itself was sometimes a source of the stigma, reinforcing popular attitudes among patients and in the community. Nurses at TTSH continued to wear face masks even though the patients' sputum was no longer infectious after several weeks of treatment, while many hospital staff (and possibly the public) continued to hold the erroneous belief that powdered milk strengthened the immune system against tuberculosis.[58] The separation of eating utensils at TTSH and among family members in the home, as advised by health visitors, also prompted some older people to believe that the disease could be spread through food.[59]

Wendy Tan, a university student who fell ill with tuberculosis at the turn of the millennium, is an example of the continuing stigma. She initially dismissed her persistent cough as 'a hundred day cough', a common Chinese belief. When she was first diagnosed with tuberculosis, she was in shock, having thought the disease to be akin to a 'soap opera' from a 'bygone era'. She felt guilty, fearing that she might infect her family and university classmates who shared the study room (they all tested positive). What surprised her, however, was the latter going to the university department to persuade her to defer her studies for a year, even though she was no longer infectious. Similarly, the young housemen at Changi Hospital, where she was warded, were reluctant to stand close to her, prompting her physician to remark, 'Would you like to step even further [away]?' The root of the stigma, Tan observed, was not ignorance of tuberculosis but partial knowledge which fuelled people's fears.[60]

Regardless of its effect on stigma, the state's health education measures had a more discernible and lasting impact in another area: on the commonplace and deeply ingrained act of public spitting. Singapore's anti-spitting campaign was an important instrument of citizenship in its sociocultural history, crossing the colonial and postcolonial years. As noted in Chapter 2, the Municipal Commission had pinpointed spitting as a cause of tuberculosis in 1907.[61] Various motivations lay behind the postcolonial version of the campaign – spitting was deemed to be anti-social and to mar Singapore's image as a modern and clean city – but another fear, underlined by the Ministry of Health in 1984, was that it could spread a large number of diseases, including tuberculosis.[62] This was a common reason for public campaigns against spitting in many countries in the time, although the risk of spreading tuberculosis from dried spit was minimal. As early as 1950, the municipal authorities had provided spittoons at some eating-places, although customers were observed to ignore them and spit onto the floor.[63]

A nascent sense of civic consciousness did arise among educated members of the public against spitting in the 1950s. A letter to the press lambasted a policeman's failure to act when a passenger spat onto the floor of the bus, despite posters expressly forbidding it in the vehicle.[64] Another writer called for the law against spitting to be more strictly enforced in cinemas.[65] Even Muslims were taken to task for spitting for the reason, as another letter claimed, that they were prohibited from swallowing their saliva during Ramadan, the holy fasting month.[66] Other newspaper articles urged the spitting ban to be extended to another act alleged to spread tuberculosis – the wanton discarding of cigarette ends onto the floor.[67]

Health education efforts against spitting assumed greater national prominence in the latter part of the decade as the constitutional and mass politics of Singapore expanded socially to the working class and ideologically to the socialist left. Anti-spitting exhortations contained the political vocabulary of self-determination and citizenship. In 1956, the *Singapore Free Press* used anti-social spitting to distinguish between 'civilised' people and 'barbarians' and its political implications for Singapore:

> We are trying to win our independence. We hope to take our place among the world's great democracies. Have we reached a state of civilisation when

we publicly show the world the wide practice of the filthy, selfish and ugly personal habit – legally banned by the greatest nations of the world?[68]

In 1958, after the People's Action Party won the City Council elections, the PAP mayor Ong Eng Guan launched a new campaign against spitting. It was, as he framed the act, not only a menace which could spread tuberculosis but also 'an anti-social habit which does not conform to the self-respect of a people approaching self-government'.[69] Ong enlisted the help of Chinese public storytellers in his Hong Lim constituency to target the Chinese population residing in Chinatown in the campaign. One of them, 'Uncle Ho', narrated to his audience how one person's incessant spitting caused a devastating epidemic which destroyed his village.[70] Ong's campaign paved the way for the PAP to redouble its efforts against spitting when the party was elected to power the following year.

The new government continued to enforce fines against offenders and decry public spitting as anti-social behaviour, while TBCU worked with the mass media to intensify its health education campaigns against the act, focusing on tuberculosis.[71] In 1961, a newspaper article warned that 'Indiscriminate spitting in crowded places in particular is a very efficient vehicle for the bacillus'.[72] There was a renewed outcry against spitting in the media in the 1980s, which both alleged the act to be a peculiarly Chinese practice and noted its disappearance among young people.[73] A 1985 article in the press lambasted spitting variously as 'Nothing short of an act of barbarism', 'Characteristic of an insensitive selfish individual', and 'Due to stupidity or lack of education'.[74] Campaigns against tuberculosis and spitting helped render Singaporeans into governable citizens.

A further STEP

In the mid-1990s, DTBC's work in the community remained centred on 'a strategy of effective treatment, early detection through contact tracing and case finding, defaulter follow-up and prevention with BCG immunisation'.[75] By 1994, the incidence of tuberculosis had fallen to 49 cases per 100,000, with 1,434 new cases reported among Singaporeans (it was not stated how many cases were of immigrants) and a mortality rate of just 3.4 per 100,000. During 1981 to 1994, the incidence had declined by an average rate of about 4 per cent per year. Most patients received short-course ambulatory therapy of six to nine months, with inpatient treatment provided only to those who were seriously ill, resistant to antibiotics or unwilling to comply with treatment. The Epidemiology Department attempted to trace the latter by phone or mail. The BCG vaccination programme remained comprehensive, with 97 per cent of all infants receiving it at or within several months of birth. The Ministry of Health's view was that 'TB is very much a disease of older people in Singapore', who were exhorted to take up the free X-ray screening.[76] It stated in 2001 that 'The leading causes of morbidity and mortality are currently the major non-communicable diseases such as cancer, coronary-heart diseases, strokes, diabetes, hypertension and injuries', with environmental and socio-economic developments helping to reduce the incidence of communicable diseases.[77]

But, as in the early 1970s, the anti-tuberculosis programme remained reflexive and concerned at the end of the millennium. The year 1997 heralded a new milestone in the history of tuberculosis control with the government's launch of the Singapore Tuberculosis Eliminatory Programme. The programme resolved to eliminate the disease within 15 years through a combination of enhanced measures: chemotherapy via DOTS, epidemiological surveillance, case-detection, early defaulter tracing, and BCG vaccination, aided by a computerised surveillance and monitoring system to track the patient's treatment to completion.[78] Explaining the rationale for STEP, the Minister for Health Yeo Cheow Tong pointed to the high incidence of tuberculosis in Singapore, numbering in the 40s and 50s per 100,000, compared to 5–10 in developed countries such as the US and Britain. The root cause, he said, was patient behaviour, with only 60 per cent of patients completing their treatment, which should ideally be at least 95 per cent, and which also compounded the likelihood of drug resistance to tuberculosis.[79]

Although spearheaded by the government, STEP was an international project in several ways. The government consulted an international advisory board that included members with substantial experience in dealing with the tuberculosis-HIV epidemic in New York City in the 1980s.[80] The connection between tuberculosis and migration, evident since the 1970s, also finally gained official attention. The Ministry of Health recognised that Singapore's heavy recruitment of migrant workers from countries with high prevalence of the disease was a real problem. However, while STEP entailed greater surveillance of short-term migrant workers, those found to have tuberculosis often did not receive treatment, as their contracts were likely terminated by their employers, which led them to be repatriated while still infected.[81] At the heart of STEP was DOTS, a global programme which was enthusiastically pushed by WHO as a universal and complete solution to tuberculosis in the mid-1990s, though there was little prior research to support its efficacy.[82]

The focus of STEP was on the patients. DOTS would be carried out at TBCU, polyclinics and, for elderly and frail patients, their homes. About two-thirds of all patients taking their medication would be supervised by nurses daily in the first two months and thrice weekly thereafter. In 1999, the proportion of TBCU's patients on DOTS almost doubled from 36 per cent in 1997 to 70 per cent. Such directed surveillance was not new; Dr Chew Chin Hin noted its roots in the supposedly 'fully-supervised' treatment of the 1970s (or Regimen 1 of the two chemotherapy courses at the time), which was usually given to patients with moderate or advanced tuberculosis, or if they had positive sputum smears.[83] Dr Goh Kee Tai had also written of a 'fully supervised intermittent outpatient treatment' in operation in the late 1970s and early 1980s as being useful for treating non-adherent patients.[84] Thus, the typical subject of STEP and DOTS was the older, sometimes non-adherent patients. Under the Infectious Diseases Act of 1976, persistent defaulters were warded at the Communicable Diseases Centre until they completed their treatment.

Although it appeared to lay the blame at the feet of the patient, non-adherence was a broad and complex phenomenon, governed by a variety of factors. Christian McMillen surmises it as primarily a systemic issue, rather than a failing of individuals, for patients' ability to complete their treatment was hampered by weaknesses in tuberculosis control and surveillance, such as their access to drugs, the length of treatment and the lack of staff.[85] In Singapore, some patients did not take their medication because they had no symptoms and thus deemed themselves to be healthy. Others stopped treatment when the symptoms began to disappear and they believed they had recovered, or when they ran into financial difficulties. According to respiratory physician Dr Cynthia Chee, it was especially hard to convince some groups of patients such as drug addicts and alcoholics to complete their treatment.[86] Some Muslim patients refused to take their medication during the holy fasting month. For many patients, swallowing a large number of sizeable pills over a long period of treatment (18 months, subsequently reduced to six) was an unpleasant experience.

The seemingly uncomplicated antibiotic treatment of tuberculosis frequently turned out to be a difficult experience when viewed from the patient's vantage point. Wendy Tan, the university student who contracted tuberculosis in 1999, had a 'very painful' experience during her six-month-long treatment. She found it difficult to take an excess of 20 pills, which were hard to swallow, at one time; she admitted, 'Many times I felt like giving up'. That she did not do so and successfully finished the course was due to her greater fear of remaining sick with tuberculosis, and to the encouragement of her mother and the doctors and nurses at Changi Hospital.[87] In the experience of Dr Edmund Monteiro, who was posted to TTSH in 1965, it was quite understandable why patients did not finish their medication, but only when doctors made an effort to establish the reasons. Some patients also suffered from side effects such as nausea and tiredness, making them feel initially worse than prior to treatment. The most common reason, in Monteiro's view, was the sheer tedium of taking the large dose of 20 pills, each with an unpleasant flavour and odour, over a long period. To do so was a remarkable feat: 'It took a little bit of courage to take those tablets for so long. But the majority of patients actually complied. That's the amazing thing.'[88]

Monteiro considered that it was thus unhelpful to simply blame the patients: one should instead listen to them to uncover the reasons, while some leeway should be allowed, such as to take the tablets after work to minimise the side effects. For Monteiro, the solution to defaulting was not strictly enforced adherence but empathy. Only then would having patients take their pills in the presence of a nurse, with coffee and biscuits, not create unwanted surveillance, but instead have a positive effect in transforming their treatment.[89]

The relationship between nurses and patients was thus key to DOTS. Dr Chee noted that observed therapy was 'not just watching someone swallow the medicine', but also required an effort to educate the nurses and to build rapport with patients to complete their treatment. Leong Chew Yin, a nurse at TBCU who could speak the major Chinese dialects and some Malay, had no problem communicating with the patients in the DOTS programme.[90] Her colleague Pushparani recalled that while some patients tried to spit out the tablets and hide them in their pocket, the nurses' patience and their creative compromises with patients could overcome the reluctance. One of her patients was given a full hour to complete her dose while combing her hair, another took the medication with bananas, and it was beneficial for some patients to take the pills together with sweets. Pushparani found DOTS to be mostly effective except in rare cases.[91]

A decade after its launch, STEP reduced the incidence of tuberculosis to the mid- to upper-30s per 100,000. As in other countries, it worked in Singapore to a degree. But the programme had yet to achieve the full eradication desired, partly because of the continuing increase in Singapore's elderly population. The rising proportion of foreigners with tuberculosis from 29 per cent in 2004 to 47 per cent in 2010, who had likely brought the disease to the city-state from countries with high prevalence, also played a role. In a 2012 study, two of TBCU's physicians, Dr Chee and Dr Wang Yee Tang warned that the repatriation of untreated foreign workers from Singapore was a short-sighted policy which would only worsen the spread of tuberculosis in the long run.[92] The island was located in a high-risk region which contributed to 29 per cent of all global cases of the disease.[93]

Tuberculosis control progressively imbued the adult population of post-war Singapore with a new ethos of citizenship. Beyond the workings of chemotherapy and radiography, the population encountered a wide array of the state's medical workers, who obtained their personal information and urged them to complete their medication, undergo X-ray screening or modify their lifestyle. The adults were also the target audience of the anti-spitting campaign, which depicted a commonplace act as anti-social, anti-national and dangerous to public health. The conference of citizenship was not always successful: some patients eschewed outpatient treatment and contact tracing, while the stigma largely persisted and in some ways worsened.

But to a large degree, the fight against a dreaded disease had taken place as an action programme with a national administrative and cultural framework, utilising a national discourse. The programme changed many of the people's attitudes and behaviour and rendered them more governable within the space of a generation. These citizen-making characteristics were also manifest in the 1997 programme to eliminate tuberculosis, which placed a strong emphasis on patients' responsibility to complete their treatment while cautioning about the entry of foreigners with tuberculosis into the country. In addition to the adults, there was one other target group of the anti-tuberculosis programme: infants and children.

Notes

1 Speech by the Minister for Health, A J Braga, at the Opening of the Institute of Health, 14 May 1958, www.nas.gov.sg/archivesonline/speeches/record-details/dc496573-bcf1-11e6-b045-0050568939ad.

2 Oral History Centre, National Archives of Singapore, Interview with N.C. Sen Gupta, Reel 1, 5 February 1999; *Straits Times*, 20 September 1976.

3 SATA, *Annual Report 1953*.

4 Oral History Centre, National Archives of Singapore, Interview with Winnie Phoon, Reel 3, 20 June 2008.

5 Medical Department, *Annual Report 1950*, p. 126.

6 Medical Department, *Annual Report 1950*, p. 126.

7 O.B. Leathart, 'The Almoner's Work in Connection with Tuberculosis in Singapore', Pan-Malayan Tuberculosis Conference, *Transactions of the First Pan-Malayan Tuberculosis Conference*, 1–4 November 1956 (Singapore: Government Printing Press, 1957), p. 1.

8 Goh Keng Swee, *Urban Incomes and Housing: A Report on the Social Survey of Singapore, 1953–54* (Singapore: Department of Social Welfare, 1956).

9 Medical Department, *Annual Report 1950*.

10 Mary Mostyn, 'B.C.G. Vaccination', *Transactions of the First Pan-Malayan Tuberculosis Conference*, p. 6.

11 Mary L. Grove-White, 'Review of the First 500 Cases in Receipt of Financial Assistance under the Tuberculosis Treatment Allowance Scheme of the Colony of Singapore', *Medical Journal of Malaya* 7 (4), June 1953.

12 E.F. Middleditch, 'The Tuberculosis Patient and his Financial Needs', *Transactions of the First Pan-Malayan Tuberculosis Conference*.

13 SWD 83/56 Tuberculosis Treatment Allowance Scheme reports, 1955–1956.

14 Middleditch, 'The Tuberculosis Patient and his Financial Needs'.

15 Speech on 'Labour and Welfare Services' by L.C. Goh, Permanent Secretary, Ministry of Labour and Welfare, 23 May 1958, www.nas.gov.sg/archivesonline/speeches/record-details/df4bf655-bcf1-11e6-b045-0050568939ad.

16 Ministry of Health, press statement, 'Anti-Tuberculosis Services', 15 March 1975, www.nas.gov.sg/archivesonline/speeches/record-details/7d2538e5-115d-11e3-83d5-0050568939ad.

17 *New Nation*, 18 February 1974.

18 Oral History Centre, National Archives of Singapore, Interview with Cecilia Nayar, Reel 3, 25 April 2000.

19 TBCU, *Brief Report for 1959 (January–October)*, p. 8.

20 TBCU, *Brief Report for 1959 (January–October)*, p. 8.

21 'Report of Dr. A. Morland on Tuberculosis in Malaya', p. 277.

22 G.P. Bardsley, 'Early Days of Treatment at SATA', in SATA, *The Royal Singapore Tuberculosis Clinic of the Singapore Anti-Tuberculosis Association* (Singapore: D. Moore, 1954), p. 58.

23 Ivy L. Foo, 'Tuberculosis Health Visiting in Singapore', *Transactions of the First Pan-Malayan Tuberculosis Conference*, p. 2.

24 Bardsley, 'Early Days of Treatment at SATA', p. 57.

25 Ministry of Health, *Annual Report 1957*.

26 In 1974, the number of routine contact examinations was reduced to two: the first following notification, and the second a year later.

27 Goh Kee Tai, *Epidemiological Surveillance of Communicable Diseases in Singapore* (Tokyo: Southeast Asian Medical Information Center, 1983).

28 TBCU, *Annual Report 1975*.

29 DIS 161/58 Memo of Australian High Commission, 'Australia Aids Singapore in Fight Against Tuberculosis'.

30 CSO TRY 2149/56 Memo from DMS to DFS, 18 June 1956.

31 TBCU, *Annual Report 1963*, p. 3.

32 Ministry of Health, *Annual Report 1960*, p. 22.

33 Oral History Centre, National Archives of Singapore, Interview with Bala Subramaniom, Reel 11, 16 October 2008.

34 Ministry of Health, *Annual Report 1965*, p. 256.
35 Ministry of Health, *Annual Report 1961*, p. 30.
36 TBCU, *Annual Report 1960*, p. 3.
37 Ministry of Health, *Annual Report 1965*, p. 254.
38 Ministry of Health, press statement, 'Anti-Tuberculosis Services', 15 March 1975.
39 TBCU, *Annual Report 1971*, p. 5; Chan Heng Chee, *The Dynamics of PAP Dominance: The PAP at the Grassroots* (Singapore: Singapore University Press, 1976).
40 TBCU, *Annual Report 1973*, p. 5.
41 Goh, *Epidemiological Surveillance of Communicable Diseases in Singapore*, p. 231.
42 *Straits Times*, 19 September 1981.
43 *Epidemiological News Bulletin*, 11 (1), January 1985, p. 5.
44 Medical Department, *Annual Report 1950*.
45 Medical Department, *Annual Report 1951*.
46 Medical Department, *Annual Report 1952*.
47 B.R. Sreenivasan, 'The Psychology of Tuberculosis', *Transactions of the First Pan-Malayan Tuberculosis Conference*, 1–4 November 1956 (Singapore: Government Printing Press, 1957), p. 10.
48 H.M. McGladdery, 'On Advising Surgical Treatment to a Patient with Pulmonary Tuberculosis', *Transactions of the First Pan-Malayan Tuberculosis Conference*, p. 15.
49 'SATA and the Un-Co-operative Patient', in SATA, *The Royal Singapore Tuberculosis Clinic*, p. 96.
50 Medical Department, *Annual Report 1954*, p. 77.
51 Speech of the Minister for Health Yong Nyuk Lin at the Opening of the Anti-Tuberculosis Week Exhibition, 7 April 1964, www.nas.gov.sg/archivesonline/speeches/record-details/78c012cb-115d-11e3-83d5-0050568939ad.
52 Kah Seng Loh, Interview with Goh Kee Tai, 16 February 2017.
53 Ministry of Health, *Annual Report 1963*, p. 2.
54 Ministry of Health, *Annual Report 1965*, p. 254.
55 *New Nation*, 29 October 1974.
56 Goh, *Epidemiological Surveillance of Communicable Diseases in Singapore*, p. 232.
57 *Epidemiological News Bulletin*, 11 (1), January 1985, p. 7.
58 Oral History Centre, National Archives of Singapore, Interview with Edmund Hugh Monteiro, Reel 3, 16 October 1997.
59 Kah Seng Loh, Interview with Leong Chew Yin, 30 March 2017; and with Pushparani, 14 March 2017.
60 Kah Seng Loh, Interview with Wendy Tan, 28 March 2017.
61 Singapore Municipality, *Administration Report 1897*, p. 85.
62 *Singapore Monitor*, 12 May 1984.
63 *Singapore Free Press*, 20 May 1950.
64 *Straits Times*, 24 September 1954.
65 *Straits Times*, 20 December 1952.
66 *Straits Times*, 25 May 1949.
67 *Singapore Free Press*, 28 February 1951.
68 *Singapore Free Press*, 25 May 1956.
69 *Straits Times*, 29 June 1958.
70 *Straits Times*, 3 August 1958.
71 *Straits Times*, 15 December 1964.
72 *Singapore Free Press*, 28 July 1961.
73 *New Nation*, 17 February 1981.
74 *Straits Times*, 5 June 1985.
75 Ministry of Health, *Annual Report 1994*, p. 35.
76 Ministry of Health, *Annual Report 1994*, p. 35.
77 Ministry of Health, *Annual Report 2001*, p. 3.
78 Ministry of Health, *Annual Report 1997*.

79 Speech by Yeo Cheow Tong, Minister for Health, at the Launch of the Singapore Tuberculosis Elimination Programme, 4 April 1997, www.nas.gov.sg/archivesonline/speeches/record-details/76487a2b-115d-11e3-83d5-0050568939ad.
80 Kah Seng Loh, Interview with Cynthia Chee, 2 March 2017.
81 Kah Seng Loh, Interview with Cynthia Chee, 2 March 2017.
82 Christian W. McMillen, *Discovering Tuberculosis: A Global History, 1900 to the Present* (New Haven, CT and London: Yale University Press, 2015).
83 Oral History Centre, National Archives of Singapore, Interview with Chew Chin Hin, Reel 6, 7 July 1999; *Epidemiological News Bulletin*, 11 (1), January 1985.
84 Goh, *Epidemiological Surveillance of Communicable Diseases in Singapore*, p. 233.
85 McMillen, *Discovering Tuberculosis*.
86 Kah Seng Loh, Interview with Cynthia Chee, 2 March 2017.
87 Kah Seng Loh, Interview with Wendy Tan, 28 March 2017.
88 Oral History Centre, National Archives of Singapore, Interview with Edmund Hugh Monteiro, Reel 3, 16 October 1997.
89 Oral History Centre, National Archives of Singapore, Interview with Edmund Hugh Monteiro, Reel 3, 16 October 1997.
90 Kah Seng Loh, Interview with Leong Chew Yin, 30 March 2017.
91 Kah Seng Loh, Interview with Pushparani, 14 March 2017.
92 Cynthia Bin-Eng Chee and Yee Tang Wang, 'TB Control in Singapore: Where Do We Go from Here?', *Singapore Medical Journal* 53 (4), 2012, pp. 236–238.
93 *Today*, 22 December 2014.

8 Newborns and children of the nation

The anti-tuberculosis programme also reached the adult population of Singapore through newborn infants and young children. The maternity hospital was an important place where parental consent was sought and given for the BCG vaccination of infants. This was key to the prevention of tuberculosis across the colonial and postcolonial periods; as two government doctors stated in 1974, 'The newborn infants have always been our first priority' in the immunisation programme.[1] Principals, teachers and other school staff likewise worked with the state's medical officers and parents to conduct X-ray screening, health checks, contact tracing, tuberculin tests, and vaccination for school students, and to implement the feeding scheme for malnourished pupils. Another feeding scheme for children with tuberculosis operated in their homes, which sought not only to improve their diet but also transform the attitudes and behaviour of their parents.

These measures not only dealt with matters of illness, sanitation and nutrition, but also had the effect of extending the influence of the state into schools and people's private lives in the urban and rural areas of Singapore. The childhood programme paralleled the control of tuberculosis among adults. As with the efforts discussed in the previous two chapters, the development of the infant and childhood anti-tuberculosis programme spanned the colonial and postcolonial periods. Both the colonial and People's Action Party governments considered and reconsidered BCG vaccination, worrying alternately about infants and older youths. The immunisation programme became increasingly reflexive under the PAP in the 1960s and 1970s, even as the campaign against tuberculosis was being won.

School X-rays, checks and contact tracing

In the 1948 Tuberculosis Policy memorandum, the British colonial government had demonstrated a deep concern with the spread of the disease among children and with their nutritional health. Admittedly, some of these supposed 'children' were older teenagers or even young adults up to 20 years of age, who were returning to school after the Second World War after their studies were interrupted by the Japanese occupation. The government surveys of school children in 1946 and thereafter appeared to confirm that the disease had become more prevalent in children during the war. This brought widespread fears of tuberculosis being transmitted among children through the use of musical wind instruments.[2]

Colonial efforts to combat child tuberculosis entered private homes as they did in the adult control programme, but they also reached a new site: namely, the school. These measures reinforced the expansion of state control over education in Singapore and in particular primary schools immediately after the war. Tuberculosis control reinforced a concerted colonial policy to reorganise education in the British image and in accordance with British values. This took place at the same time the Medical Department's almoners and health visitors were appearing in people's homes and influencing what transpired within.

A binary education system of sorts existed in pre-war Singapore: most government and government-aided schools taught in English, while numerous privately run and funded schools stood beyond the purview of the colonial regime, catering to various ethnic groups and teaching in their vernaculars. After the war, as they did with the Medical Plan, the British looked to reshape the education system as part of the decolonisation of Singapore. In 1947, the Ten-Year Education Plan was launched, aiming at 'fostering and extending the capacity for self-government, and the ideal of civic loyalty and responsibility', and more generally at meeting the development needs of postcolonial Singapore.[3] The focus of the plan was on primary education, with free primary education provided for both sexes, and ten new primary schools to be built yearly (this was increased to 18 in 1950). In 1954, the colonial government further extended its influence over education by substantially increasing grants-in-aid to the independent vernacular schools. By 1959, there were only 75 unaided primary schools in Singapore – compared to 272 government and 257 aided primary schools – in which were enrolled less than 5 per cent of the primary student population.

Sanitary and public health measures to safeguard the health of school children and staff were less visible but no less instrumental means to remake and transform the primary schools of Singapore throughout the 1950s. Schools that were applying to register with the state to qualify for official grants had to comply with minimum sanitary requirements. The control of tuberculosis was one such criterion, with how 'stress is always placed on the provision of ample playing fields and satisfactory tiffin-sheds, adequate sanitary arrangements, good water supply and the proper lighting and ventilation of classrooms'.[4] Like people's inner city homes, many Chinese private schools were small and overcrowded. Throughout the 1950s, as more schools registered and qualified for state support, their sanitary conditions also improved.

Conversely, the Medical Department disapproved of unlicensed hawkers commonly found peddling their food outside the school gates, regarding it as lacking in hygiene and nutrition. The department preferred the regulated school tuckshops operated by state-licensed vendors, similar to the hawker centres subsequently established in Singapore's public housing estates. But the British were unable to disperse the itinerant vendors found outside school premises.

The state's medical officers appeared increasingly in the schools to examine the health of pupils. In 1948, 34,177 such examinations were conducted, mostly of new entrants and school leavers – a rise of more than half from the previous year. But the response differed among school types: while government and aided schools generally accepted the inspections, many unaided vernacular schools resented them as an unwanted presence and intervention. The following year, the department conducted examinations at nearly all government, aided and Malay schools, though they were able to inspect Chinese aided schools only every two to three years due to their large numbers and the shortage of staff. Most students examined were found to be in good or improving health, with 12 per cent that year deemed to have signs of malnutrition compared to 16.15 per cent the year before. Yet the examinations continued, rising in number in the following decade. In 1958, 74,058 pupils were examined, or close to a quarter of the student population.

As with the adult tuberculosis control programme, the surveillance of the schools had a spatial dimension, enabling the expansion of government influence into the outlying and rural areas of Singapore. This gave the state access to greater numbers of children from the lower-income group, who were deemed to have relatively poor physical health and to be more susceptible to tuberculosis. In 1956, the Medical Department made visits to 459 out of a total of 504 government and aided schools. Of these schools, 252 were located in the city while 190 were in the rural area. Parents, too, were asked to participate in these programmes. In 1949, the Medical Department asked that a parent accompany the child during the health examination, but this request was seldom acceded to, and often only in the examination of girls.

The anti-tuberculosis school programme formed an important part of this expansion of public education and school health. It came under the purview of the school health service of the Medical Department (and subsequently the Ministry of Health) until 1958, when the service was transferred to the Assistant Director of Medical Services (Tuberculosis), who oversaw both the preventive and treatment aspects of the disease. Although the service was also concerned with the dental health, physical defects and other infectious diseases of children, its primary role, performed by the Control of Tuberculosis section, was to tackle tuberculosis among school students, teachers and hawkers. The section's work was substantial and wide-ranging, including the medical examinations and X-ray screening of students, teachers and hawkers. This involved case-finding work in homes, schools and hospitals, namely tracing and examining school children of parents who had tuberculosis, and the contacts of students with the disease. From 1951, as we will see in what follows, the section also carried out the massive work of BCG vaccination for primary school pupils. It also ran school clinics for pupils with tuberculosis, thus extending outpatient treatment to school children.

As with its extensive use for adults, chest X-ray screening was a vital preventive measure to detect tuberculosis among school children and staff. In 1949, 3,174 students and 1,520 teachers were examined for the disease, as the Medical Department ruled that teachers had to be X-rayed before being employed by a school if they had not done so the previous year. The number of individuals detected with tuberculosis was fairly small but not insignificant in light of the possible infection of children. In 1952, of the 2,937 teachers screened, 50 were found to have active tuberculosis, and ten of 325 school hawkers – a significant proportion in both instances. In 1956, when 13,394 school children and school staff were screened, 264 children and one teacher were discovered with active tuberculosis, while 26 children and 50 staff persons had the chronic form of the disease. This prompted the Medical Department to warn, 'A whole school population is at risk from one infectious case in their midst'.[5] The department was further worried that while older teachers, being used to the medical checks, were more willing to undergo the procedure, there was some resistance from younger teachers.[6] There was at least little trouble with hawkers and clerical staff, whose screening was usually supported by the principal. In 1959, visits to secondary schools in urban and rural areas uncovered a surprising number of active pulmonary tuberculosis cases – 19 out of 1,701 students tested.[7]

Pupils detected with tuberculosis were sent to TTSH for assessment and treatment, and their contacts were also investigated and screened. The hospital had opened special wards for children, though in cases where circumstances and finances permitted, some children were treated at home. In 1948, a health nurse in the Medical Department was placed in charge of children with tuberculosis, attending to a wide range of matters: ensuring their regular attendance at the school clinics; accompanying them to a hospital or clinic; and checking on their diet and medication. This was culturally sophisticated work as with adult patients, which required, as the department stated, 'Tact and a good knowledge of psychology' in addition to medical expertise.[8] In 1954, the department's personnel made 947 and 627 visits to children with tuberculosis in homes and schools, respectively. They reported that defaulting of treatment among children had virtually been eliminated as a result of the visits.

The schools were connected to contact clinics for children of school-going age. As the Medical Department reported, these school clinics went beyond the function of a standard school clinic, and were in fact 'outpatient departments for children of school age'.[9] These outpatient nodes comprised the main clinic in the city, located at North Canal Road, and smaller clinics outside the urban boundary at Paya Lebar, Kallang and Bukit Timah. The North Canal Road clinic, as the department described in 1951, was 'a small cramped building liable to be flooded when heavy rain coincides with a high tide', operating on Monday, Friday and Saturday mornings, and in the afternoon on all days except Saturday and Sunday. It was inadequate to meet the needs of the large numbers of school students who came to its doors.[10] In 1958, the clinic was replaced by a new one located within the newly established Institute of Health at Outram Road. The institute, which also housed the School Tuberculosis Service, served as the primary centre for public health work and teaching. The subsidiary school clinics outside the city only opened on some afternoons of the week. From 1954 to 1958, the number of attendees at the clinics doubled from 48,414 to 95,106. In addition, the Medical Department operated a travelling dispensary to treat minor ailments of pupils in schools and to follow up on BCG vaccination.

The considerable colonial efforts laid the foundation for the child tuberculosis programme after 1959, which remained a primary concern for the PAP government. The School Tuberculosis Service, now under TBCU, carried out X-ray screening for primary school entrants and primary and secondary school leavers. As before, it also conducted case-finding for young contacts of tuberculosis patients, including those not enrolled in school or of school-going age, and for contacts of students who had tuberculosis. The four school contact clinics in Singapore continued to play important roles in examining and monitoring the child population. These were the main clinic located at the Institute of Health, which served the city, and three outpatient clinics in the urban and rural areas, namely, the Paya Lebar, Kallang and Bukit Timah clinics.

The PAP similarly maintained and expanded the other instruments of the child tuberculosis programme. A comprehensive system of health checks continued to be carried out in both urban and rural schools, including the mass tuberculin testing and BCG vaccination of primary school children, and the feeding programme, which are discussed in the following sections. Primary school entrants and leavers in the Primary One and Six classes respectively still received, subject to parental consent, the Heaf Multiple Puncture Tuberculin test, which was employed up to 1965, and thereafter the Mantoux test. Students with negative reactions were given BCG vaccination while those with strongly positive reactions to the tests were X-rayed for active tuberculosis. Secondary and pre-university school leavers also underwent chest X-rays, while older boys were screened at their medical examination for National Service by the Central Manpower Base.

There were a number of minor administrative changes to the screening regime after 1959. In 1961, the almoner's office at the Institute of Health took over cases previously handled by TTSH, most of whom were children being assessed for the Tuberculosis Treatment Allowance Scheme or children with tuberculosis on the special feeding scheme. The following year, TBCU assumed the routine X-ray screening of school teachers, hawkers and other staff every two years. The school health system as a whole did not change substantially. The Ministry of Health expressed its satisfaction in 1965 that 'The notification rate, and tuberculin rate amongst primary school children have dropped slowly but surely'.[11] The inspections and immunisations helped to eradicate tuberculosis among children in the 1960s. They were not let up: around two-thirds of new tuberculosis cases in 1975 were detected during the routine school screening.

BCG vaccination, reflection and reform

The 1950s were significant for the history of BCG vaccination in Singapore, heralding the end of the initial colonial reservation towards it, followed by an expanding immunisation programme for infants and children. In his visit to Singapore and Malaya in 1948, Andrew Morland had proposed an extensive trial of the vaccine. He maintained that the vaccine could be safely used by susceptible groups, such as hospital nurses, medical students, children of tuberculosis families, and military recruits, and also the large Chinese population dwelling in shophouses. Malay children, he surmised, had lesser likelihood of tuberculosis infection.[12] Encouraged by Morland's pronouncements, local doctor Chen Su Lan and G.H. Garlick, Director of SATA's Royal Singapore Tuberculosis Clinic, subsequently urged the use of BCG as a low-cost way to combat tuberculosis, while a *Straits Times* commentary even talked up the idea of vaccinating the entire population of Singapore.[13]

In 1949, in a project jointly organised by the Swedish Red Cross and World Health Organization, Dr Johannes Holm, the Technical Director of the International Tuberculosis campaign, visited Singapore to offer his advice on BCG vaccination. He proposed the use of a wet form of the vaccine for infants, children and susceptible groups such as contacts of those with tuberculosis and medical personnel working on the disease; like Morland, he emphasised it as being harmless to people. Holm pointed out that in Denmark, which had the lowest tuberculosis death rates in the world, family members of someone with the disease were also given the vaccine if they tested negative for the disease.[14] That year, the Medical Department accepted that BCG 'must soon take its part in the Singapore picture'.[15]

In 1951, the World Health Organization and United Nations International Children's Emergency Fund sent a team of experts led by Dr Arne Buus-Hansen to Singapore to launch a pilot vaccination programme. The team trained personnel from the school health service, Kandang Kerbau Hospital, the main government maternity and children's hospital in Singapore, and maternity and child welfare centres in immunisation work. The service's BCG team proceeded to carry out tuberculin testing and vaccination of newborn infants and primary school students on a voluntary basis, after parental consent had been obtained. This excluded adults of 30 years of age and above, who would usually be tuberculin-positive.

Although there were ongoing criticisms of BCG vaccination globally, in Singapore the programme proceeded smoothly, beginning with the pupils of the Convent of the Holy Infant Jesus, a government-aided missionary school. The authorities assured the public that there was no need to be concerned about the large percentage – 70 per cent – of positive tuberculin test results, especially among pupils of schools in the crowded inner city of Chinatown.[16] Of the 49,067 tests carried out that year, there were 18,784 vaccinations. Parental consent for children to be screened and vaccinated was obtained in virtually all cases, as was the cooperation of schools.

Immunisation proceeded apace. For the next few years, BCG was given not only to primary but also to older secondary students. In 1954, the BCG team visited 63 schools, of which 19 were government, eight Malay, 27 Chinese, two Tamil, and five private schools. It tested over 9,000 children above the age of ten using the Mantoux method, following which one-third were vaccinated. By 1956, the team had visited nearly all primary and secondary schools in Singapore. The following year saw, however, a major change to the immunisation programme. On the advice of the Assistant Director of Medical Services (Tuberculosis), the

Mantoux test was replaced by the Multiple Puncture test which, being simple and painless, was suitable to be used initially on younger children without making them unduly apprehensive at vaccination. The result was that immunisation was reduced in scope to only Primary I entrants, excluding secondary school students.

The BCG team also made use of the school holidays to test and immunise children who were not enrolled in schools, but who were living in institutions supervised by the Social Welfare Department, and in the orphanages and homes of the Salvation Army. The numbers of tuberculin tests and BCG vaccinations performed in schools during 1957 to 1959 were 113,676 and 49,468 respectively.

Unlike the immunisation of pupils, the programme for newborn infants encountered some resistance from parents and was more uneven. In 1951, compared to the vaccination of school children, only half of the mothers at KKH gave consent for their infants to be immunised.[17] The following year, the Medical Department appeared to accept the recommendation of Professor Frederick Heaf that immunisation should not be given to infants and toddlers upwards in age but should concentrate on older children from age 20 and less, and on the susceptible groups. This led to the testing and vaccination of secondary school students as well as primary school students discussed earlier.

Despite this, the vaccination of infants did not end as a priority and the basic desire to protect them from naturally occurring infection at birth remained. At the Pan-Malayan Tuberculosis Conference held in Singapore in 1956, Dr Mary Mostyn, a chest physician at the Royal Singapore Tuberculosis Clinic, stated that the immunisation of infants was safe, crucial and effective.[18] The following year, in addition to the vaccination of infants under a month old at KKH, which served the city area, the BCG team launched a pilot infant immunisation project at two child health centres in the rural area, expanding the reach of vaccination work. This proved to be a success – 'the response from the parents was most encouraging'[19] – and the scheme was extended the following year to all 20 rural centres in Singapore, as well as to the three centres within city limits. Conversely, the government turned away from Heaf's recommendation that year and ceased vaccinating secondary school students. TBCU further stepped up the infant vaccination campaign, which in 1959 covered 28,283 out of 29,269 infants discharged from KKH. Attempts to vaccinate infants outside the hospital resulted in more objections from parents: in that year, only 5,936 of 14,773 infants born outside the hospital but within the city, and an even smaller proportion of 4,078 of 15,359 infants born in the rural area, were vaccinated.

After 1959, the Ministry of Health's school health service continued to conduct chest X-ray screening of school entrants and leavers, and biannual X-ray checks for school staff. In addition, they also investigated student contacts of notified cases and pupils suspected of tuberculosis found during the routine visits to schools.[20] Schools, hospitals and clinics remained key sites of tuberculosis prevention and case-finding for infants and children. The School Tuberculosis Clinic diagnosed and treated active tuberculosis cases from schools, carried out tuberculin tests and gave BCG to pupils with negative reactions. The PAP government received international assistance in the BCG programme as the colonial administration previously did. In 1960, under the auspices of the Colombo Plan for regional technical assistance, Singapore received free vaccines and tuberculin derivatives for the Heaf Multiple Puncture test from Australia, which were previously purchased from Britain and Denmark. In 1963, WHO and UNICEF provided equipment and transport for mobile vaccination and X-ray screening to Singapore.

The lynchpin of the postcolonial childhood tuberculosis programme remained the BCG vaccine, but its history in Singapore turned interesting as further refinements were made. The vaccination regimen remained comprehensive, and BCG continued to be given at a range of locales: initially to children in primary schools and institutional homes, and to infants at KKH and maternity and child clinics in the urban and rural areas, which were served by mobile immunisation teams. In 1960, the BCG dosage for infants was reduced to 0.05 cc following reports of lymphadenitis reactions. As before, primary school entrants were given BCG if they had not been vaccinated at birth or if they tested tuberculin-negative. The number of tuberculin tests for primary school students rose from 61,436 in 1959 to 90,465 in 1965.

The school health service covered an average of 450 schools per year between 1961 and 1966. By medium of instruction, these institutions comprised 235 Chinese schools (52 per cent of total), 190 English schools (42 per cent), 37 Malay schools (8 per cent), and 15 Indian schools (3 per cent). The figures highlighted a significant departure from the colonial period, with Chinese-medium schools receiving greater emphasis from the state health service than before. In particular, they underlined the postcolonial government's growing control over Chinese-medium education. As early as 1960, 96 per cent of the student population in Singapore were already enrolled in government and aided schools.

But though much of the groundwork had already been laid, the 1960s were still historically significant years for the immunisation programme. Another big change was attempted in 1961, with the government making a bold move to prepare legislation for the compulsory vaccination for tuberculosis of not only newborn infants and primary school entrants, but also primary school leavers and their contacts, and also the compulsory vaccination for another infectious disease, diphtheria.[21] The reasoning for tuberculosis was that although the BCG vaccine was safe to use and effective in reducing the disease among the young, only 60 per cent of parents were assenting to its use on their infants.[22] In 1960, the Ministry of Health reported a 'slowly increasing response' to the BCG vaccination of newborn infants.[23]

The mandatory law would have taken the colonial control programme a major step further, but it was rather surprisingly dropped the following year. The ministry briefly explained in 1962, 'the response [rate] was satisfactory and it was decided to carry on as it is, i.e. on a voluntary basis and parental consent being still necessary before B.C.G. is given'.[24] This is interesting when compared to the similar legislation for compulsory vaccination against diphtheria, which was passed. Part of the reason may have been SATA's opposition to compulsory vaccination, as it later stated in its 1979 annual report.[25]

The absence of compulsory legislation for BCG did not end the development of the infant immunisation programme, which continued to expand within the framework of parental consent. In 1960, the largest number of infant immunisations – 33,109 – took place at KKH, with the response rate rising from 94 per cent in 1958 to 99 per cent.[26] Thereafter, as the ministry highlighted in its reports, the percentages of infants who received BCG remained high, with over 90 per cent of all babies born in KKH vaccinated in the 1960s except in 1965. As Dr Chew Chin Hin, who joined TTSH in 1957, observed, KKH was a 'captive area' for giving the BCG vaccine to infants.[27] This, combined with health screening in schools, almost entirely eradicated tuberculosis among infants and young children.

There was still parental resistance to BCG vaccination of infants in the late-colonial period, which was manifest outside of KKH in the outlying areas of Singapore. In 1962, a year after the proposed mandatory law was shelved, the Ministry of Health stipulated that a much larger number of babies up to two months old (instead of one month) would be vaccinated. This would include infants who were not immunised at birth and was ostensibly a response to continuing parental resistance outside the hospital. Infants born outside KKH still had lower rates of vaccination throughout the 1960s. In 1961, only half of the 12,891 babies born in the city but outside KKH and 10,460 infants born in the rural area were immunised. There are no similar statistical breakdowns provided in subsequent years, although in 1968 TBCU declared that 'B.C.G. vaccination coverage amongst new born infants has increased year by year'.[28] In 1965, the vaccination programme had also been extended to two islands off the Singapore main island, with 219 infants immunised on those islands.

Indeed the national infant coverage rose dramatically from 36.7 per cent in 1957 to 92.1 per cent in 1972. Between 1960 and 1970, half a million newborn infants in Singapore received the BCG vaccine. In 1975, nearly four-fifths of 28,277 newborn infants were vaccinated in government hospitals, with most of the remainder receiving BCG at the maternal and child clinics and a small but unknown number from general practitioners. The trend shifted in the 1980s. The percentage of infant vaccinations in government hospitals fell to 73.6 per cent in 1984, though this was due to growing numbers of babies being delivered in private hospitals, where they were probably immunised, but for which the government did not have statistics. There was also a growing preference among parents for infants born outside KKH to be immunised by general practitioners.[29]

The BCG programme for school children continued to undergo reflection, adjustment and reform as it did in the colonial period. In 1965, the PAP government decided to extend voluntary BCG immunisation to secondary school leavers, who would be X-rayed regardless of the tuberculin test result, and those with weak/negative reactions to the tuberculin test were offered BCG immunisation. The vaccination of secondary school students was not new, having been practised by the previous administration for a brief period between 1951 and 1956. But the PAP government's interest in secondary school students grew with tuberculosis declining among infants and younger children and becoming relatively more prevalent among older children at this time.[30] Similarly, in contact tracing, all contacts above 15 underwent X-ray regardless of tuberculin sensitivity, and a second vaccination was given to children ten years old and above with a weak or negative reaction to the tuberculin test.[31] This was within the maximum of two BCG vaccinations per person.

Thus, even as the incidence of tuberculosis declined in the 1960s, the BCG programme in schools expanded. The inclusion of secondary school leavers increased the number of school vaccinations to over 120,000 over the next two years. In 1968, TBCU conducted tuberculin tests and BCG vaccination in all government and aided primary and secondary schools in Singapore, save for a few primary schools located on the outlying islands. In 1969, two further modifications were made: the criterion of five millimetres for determining a positive tuberculin reaction was doubled to ten millimetres, while primary school entrants with a negative reaction would no longer receive a second vaccination as before, as it was felt that the BCG vaccine provided adequate protection for over a decade.[32] Between 1960 and 1968, the number of tuberculin tests for children jumped from 69,339 to a high of 120,819 before falling. A total of 420,000 school children were immunised between 1960 and 1970. In 1972, 90 per cent of primary school students and 75 per cent of secondary school students received the vaccine.[33]

In 1973, TBCU found a new problem: there was a 'marked reduction' in tuberculosis among primary school children over the past six years, but this was not the case with secondary school students.[34] It was concerned about the relatively high incidence of the disease among adolescents and young adults in the 15 to 24 years age group. Dr Ng Yook Kim, the head of the Department of Tuberculosis Control, offered several possible explanations: immunisation had been less comprehensive for this age group in the past when they were younger, they were exposed to the bacillus while at work, they consumed less nutritious food, or breathed re-circulated air in air-conditioned rooms where the bacillus was present, while female adolescents were more susceptible due to anaemia from heavy menstruation.[35] The Minister for Health Chua Sian Chin similarly highlighted the risk among youths, warning that the overall incidence of tuberculosis in Singapore was still ten times that of the US and Europe.[36]

In response, DTBC carried out a feasibility study in 1976 for direct BCG revaccination of primary school leavers, who would be immunised a second time without a tuberculin test. Such revaccination was an existing practice in Singapore; in 1968, out of 115,054 immunisations, nearly a fifth were revaccinations. But it was now proposed with some urgency by the Tuberculosis Advisory Council in response to what DTBC deemed to be 'the relatively high tuberculosis incidence in the young adult age group 15–24 repeatedly reported annually by the central tuberculosis registry', with an intention to introduce a new vaccination schedule.[37]

The crux of the matter was, as the Expert Committee on the Immunisation Programmes in Singapore warned, the apparent weakening effect of BCG vaccination in infants after a period of ten to 15 years, reversing the stance taken in

1969.[38] Despite this, direct revaccination did not become policy. The proposal contradicted WHO's position that revaccination was of limited use for older adolescents, there being no conclusive evidence of it conferring additional protection against tuberculosis, while it was also costly, and the main strategy should remain the immunisation of infants in their first year.[39] In 1980, the government appeared confident that 'There is no evidence to support the suggestion of an increasing tuberculosis infection transmission in Singapore'.[40]

While there was no rigorous controlled study of its impact, a study in 1974 maintained that BCG vaccination contributed substantially to the decline in tuberculosis morbidity and mortality in Singapore over 20 years. A little over one million vaccinations were performed during 1957 to 1972, or nearly half of Singapore's total population. Of these, 680,098 vaccinations were given to infants, 512,797 to school students, and 49,417 to contacts of tuberculosis patients. In 1972 alone, the incidence rate was five and 34 per 100,000 for immunised primary and secondary school students, compared to 37 and 127 respectively for the non-immunised group. In 1956, the death rate for respiratory tuberculosis had been four per 100,000 for children aged four and below, 0.5 for those aged from five to nine years, and 0.8 for those aged from ten to 14. By contrast, in 1971 there were no such deaths for children under ten at all, while the mortality rate for those aged from ten to 14 fell to 0.3. The study urged that the BCG programme 'should be continued and the high level of coverage should be maintained and raised further, in order to bring tuberculosis under control in Singapore in the not too distant future'.[41]

Despite its concerns and reflections, the PAP government's basic belief in BCG vaccination did not waver. In the late 1970s, the official view was that 'the probability of developing TB is reduced three and a half times with BCG vaccination'.[42] This stance did not change in 1980 when a major study of the vaccine's use in India, supported by the WHO, suggested otherwise. In response to contradictory evidence, Dr N.C. Sen Gupta, the Medical Director of SATA, pointed out that, among other problems, the Indian study did not cover infants and thus did not apply to Singapore.[43] In 1985, the government stated that as a preventive measure, BCG vaccination remained essential for the effective control of tuberculosis, especially as

> Singapore is a densely populated international port with a large number of visitors, of whom many come from countries in the region which have high tuberculosis prevalence. Singaporeans are therefore constantly being exposed to the risk of infection. Even if it is possible to formulate a fully effective local control programme, the source of infection cannot by completely controlled.[44]

The average Singaporean child born in the 1980s and 1990s was likely to receive BCG vaccination within two months of birth and possibly a second dose in primary or secondary school. In 1993, 98 per cent of newborn infants were immunised, more than half of them in private hospitals; so were 76 per cent of primary school leavers at about 11 years of age, with only very small numbers of vaccines given to primary school entrants and secondary school leavers with no previous BCG vaccination or scars.

Feeding the child

The child-feeding schemes of post-war Singapore were another prominent way to prevent tuberculosis by tackling the widespread malnourishment among young children. The schemes began originally as an emergency measure organised by the British military administration after the war to deal with the adverse health effects of the Japanese Occupation on children in schools and Welfare Centres. They evolved into permanent programmes and a staple of the activist work of the colonial and subsequently PAP government. The feeding schemes constituted the intervention of the School Health Service and Social Welfare Department into areas of private life where the role of parents was no longer considered to be adequate.

The key instrument of the feeding schemes was the scientific theory of nutrition, implemented by the Nutrition Unit of the King Edward VII College of Medicine (and from 1949 the University of Malaya). The nutritional scheme sought to bolster a person's biological resistance to tuberculosis, albeit through a Western-based diet. In post-war Singapore, the colonial administration provided high-protein meals comprising meat or fish, eggs, milk, and peanuts to several groups of adults, including malnourished persons, persons suffering from tuberculosis and expectant mothers. But the focus of the feeding programme was young children. In 1948, as the Medical Department noted in explaining its domiciliary relief scheme for children recovering from tuberculosis after their discharge from hospital, 'The great majority of parents are not financially able to provide the extras required such as eggs, or vitaminized margarine, fresh fruit, fresh vegetables and milk'.[45]

The child-feeding schemes received substantial and sustained support from the British colonial government. As early as 1946, nearly 10,000 students in 33 schools received 1.2 million meals under three protein-based feeding schemes, which were popular and successful.[46] Although these schemes ceased that year due to a shortage of funds, the government decided to continue feeding pre-school children in the two to six years age group, who were thought to be particularly vulnerable to disease. In the following year, 4,500 children from this age group received rice, meat or fish, fruit, and milk at designated feeding centres (later renamed Children's Social Centres) run by the Social Welfare Department, while free milk was also dispensed to infants and children at the maternity and child welfare clinics located in the rural area.

This feeding scheme reportedly helped three-fifths of the children gain weight and was expanded to older children the following year. It was also extended to children who were contacts of tuberculosis patients (where all children in the household were put on the scheme as a precaution, and also given BCG vaccination), to apparently healthy children who were generally malnourished, and to children suffering from leprosy. In 1948, the government claimed an improvement in the general health of school children, with the percentage of undernourished children falling by nearly half from the previous year, and further declining thereafter.

After 1948, following the formulation of a strong anti-tuberculosis policy by the colonial regime, the child nutrition and feeding programme expanded in scope. In 1950, free skimmed milk provided by the United Nations Children's Emergency Fund was made available in Singapore. This broadened the child-feeding programme, with skimmed milk being a superior substitute to dried whole milk, and usually mixed with Ovaltine or Milo (milk flavouring products containing malt extract and cocoa) to make it more palatable to local children. The drinking of free skimmed milk was widely implemented in schools in the 1950s; it was a, sometimes unwanted, rite of passage for young pupils, as teachers simply picked thin, undernourished-looking children to drink the milk during recess. The number of pupils taking free skimmed milk rose throughout the 1950s; in 1959, nearly 26,000 school students drank it.

The Medical Department had another feeding scheme that was more specific and targeted. This was for children from low-income families who had a minor tuberculosis infection (also known as an active primary complex). The scheme was home-based since most children with tuberculosis were treated as out-patients due to the shortage of hospital beds, and usually lasted six months. The children received a free ration high in protein every fortnight prescribed by a dietician at the University of Malaya, comprising 1 lb. full cream powdered milk, ½ lb. vitaminised skimmed milk powder and ¼ lb. Ovaltine for flavouring (with all three items mixed before distribution), ½ lb. fresh butter, six fresh eggs, six oranges, and 1 lb. of shelled peanuts or groundnuts. The children's parents collected the rations from feeding clinics every Thursday morning at the North Canal Road Clinic, when it operated its tuberculosis service, and at the Paya Lebar rural dispensary in the eastern part of Singapore. Unlike the general feeding programme for malnourished children, the domiciliary feeding scheme was small, originally involving about 100 children before rising to over 200 in 1951.

The efficacy of both feeding schemes is uncertain. The Medical Department assessed the children every third week and lauded its role in improving child nutrition and health. In 1949, 46 per cent of the beneficiaries ostensibly showed 'obvious and satisfactory improvement', and other favourable reports followed suit in the 1950s.[47] However, the department also expressed doubts on the scientific verifiability of its own claims, acknowledging that a child's weight gain could be due to his or her normal growth rather than to the diet. In 1951, it admitted that 'there is insufficient data available as to the normal weightage curves of Asian children in normal health and with adequate food', adding that the problem also existed in the West.[48] Nevertheless, the scheme continued, if only because, as the health authorities conceded, the alternative course of action – to build more hospital accommodation for children – was delayed in the early 1950s.[49]

The feeding schemes were also somewhat mitigated by prevailing socio-economic circumstances in Singapore, particularly among working-class families. Despite periodic checks by nurses and health visitors, the Medical Department found that the rations were sometimes shared with other children in the family due to general food needs. There was also the likelihood that the female child was unable to obtain adequate rest at home, as she was often made to carry out household tasks or care for her siblings. At least the Medical Department did not find evidence of the rations being sold for income, as it feared. For these reasons, the department proposed establishing a sizeable convalescent home for children with primary complex tuberculosis, located 'preferably at the sea-side where children can rest and play under guided supervision' as 'In most homes there is little or no fresh air and no privacy, and no separate sleeping apartment for the child'. But apparently such a home was never built, likely because the department was unwilling or unable to bear the cost, calling instead for community support and financing.[50]

The twin feeding schemes continued into the postcolonial period. Free skimmed milk from UNICEF provided by the Social Welfare Department was still given to school children, deemed to be undernourished, by school health officers from the Ministry of Health. This scheme encountered some initial difficulty and was suspended in 1960, as a number of school principals were reportedly failing to distribute the milk to their pupils. But this was found to be because of the milk's unpleasant odour, reminiscent of the situation in the 1950s. The principals were given a small budget to purchase sugar or a flavouring agent such as Milo to make skimmed milk more palatable. This turned the scheme into a success, with milk given to over 20,000 pupils in 1961 and to larger numbers in subsequent years.

In the smaller scheme, children with tuberculosis lesions, typically from low-income families, received free fortnightly rations of milk and meat as prescribed by the dietician of the University of Malaya (renamed the University of Singapore in 1962). The number of students and the budget committed to this scheme remained fairly small, with a sum of $20,000 allocated and 306 new students added to a total of just over 1,000 children in 1960. More telling was the fact that, despite the onset of postcolonial governance, the diet remained unchanged from the colonial version and was decidedly Western-centric. In a way, the postcolonial government rather over-zealously pursued the benign effects of Western medicine. In the 1970s, the government gave malnourished pupils a Wheat-Soya-Blend (WSB) diet proposed by the United Nations World Food Programme, which was an 'enriched high protein-calorie food supplement together with dried skimmed milk'.[51]

The Singapore state's policy towards tuberculosis among infants and children was as much a feat of social engineering as it was for the adult population; it involved adults as parents, principals and teachers. In addition to the Tan Tock Seng and Kandang Kerbau government hospitals, other places such as schools, clinics and homes became important sites where the attitudes and behaviour of adults were being transformed. The X-ray screening, health checks, BCG vaccination, and feeding schemes gave the government influence over both the education system and the role of parents in the care of their infants and children, although the reach was never complete.

Of particular note was the continuing reform of the immunisation regimen through the 1950s and even after 1959. Was it more useful to vaccinate infants or older adolescents? The Tuberculosis Control Unit in particular worked closely with the government to identify gaps or flaws in the vaccination programme. The prevention of tuberculosis among infants and children thus played a role in social-ising adults into citizens. Not all of the official plans were implemented: though the programme was extended to secondary school leavers in 1965, the proposed compulsory immunisation and direct revaccination of primary school leavers did not materialise. But the reflexive BCG programme between the 1950s and 1970s helped define and reinforce the government's close management of the state and society of Singapore through tuberculosis control.

Notes

1 C.H. Chew and P.Y. Hu, 'BCG Programme in the Republic of Singapore', *Singapore Medical Journal* 15 (4), December 1974, p. 242.
2 Oral History Centre, National Archives of Singapore, Interview with Paul Abisheganaden, Reel 18, 17 June 1994.
3 Education Department, *Annual Report 1954*, p. 11.
4 Medical Department, *Annual Report 1953*, p. 70.
5 Ministry of Health, *Annual Report 1956*, p. 11.
6 Ministry of Health, *Annual Report 1957*.
7 TBCU, *Report for 1959*.
8 Medical Department, *Annual Report 1948*, p. 53.
9 Medical Department, *Annual Report 1951*, p. 87.
10 Medical Department, *Annual Report 1951*, p. 87.
11 Ministry of Health, *Annual Report 1965*, p. 250.
12 'Report of Dr. A. Morland on Tuberculosis in Malaya', *The Medical Journal of Malaya* 4 (4), June 1950.
13 *Straits Times*, 10 July 1950; *Straits Times*, 24 April 1950.
14 *Straits Times*, 2 May 1949.
15 Medical Department, *Annual Report 1949*, p. 96.
16 *Straits Times*, 15 September 1951.
17 Mary Mostyn, 'B.C.G. Vaccination', Pan-Malayan Tuberculosis Conference, *Transactions of the First Pan-Malayan Tuberculosis Conference* 1–4 November 1956 (Singapore: Government Printing Press, 1957).
18 Mostyn, 'B.C.G. Vaccination'.
19 Ministry of Health, *Annual Report 1958*, p. 64.
20 Ministry of Health, *Annual Report 1966*.
21 TBCU, *Annual Report January to October 1959*.
22 *Singapore Free Press*, 25 July 1960.
23 Ministry of Health, *Annual Report 1960*, p. 34.
24 Ministry of Health, *Annual Report 1962*, p. 123.
25 SATA, *32nd Annual Report for the Year ended December 1979*.
26 Ministry of Health, *Annual Report 1960*, p. 34.
27 Kah Seng Loh, Interview with Chew Chin Hin, 8 March 2017.
28 TBCU, *Annual Report 1968*, p. 7.
29 *Epidemiological News Bulletin*, 5 (11), November 1979.
30 Ministry of Health, *Annual Report 1960*.
31 As before, younger contacts below 15 with a strong tuberculin reaction were given chest X-rays while negative reactors were given BCG.

32 Chew and Hu, 'BCG Programme in the Republic of Singapore'.
33 Chew and Hu, 'BCG Programme in the Republic of Singapore', pp. 241–245.
34 TBCU, *Annual Report 1973*, p. 18.
35 *Straits Times*, 11 April 1975.
36 *Straits Times*, 8 April 1975.
37 TBCU, *Annual Report 1976*, p. 1.
38 *New Nation*, 20 August 1975.
39 WHO, 'WHO Statement on BCG Revaccination for the Prevention of Tuberculosis', *Bulletin of the World Health Organization* 73 (6), 1995, pp. 805–810.
40 *Epidemiological News Bulletin*, 6 (9), September 1980, p. 37.
41 Chew and Hu, 'BCG Programme in the Republic of Singapore', p. 244.
42 Goh Kee Tai, *Epidemiological Surveillance of Communicable Diseases in Singapore* (Tokyo: Southeast Asian Medical Information Center, 1983), p. 232.
43 *Straits Times*, 6 August 1980.
44 *Epidemiological News Bulletin*, 11 (1), January 1985, p. 5.
45 Medical Department, *Annual Report 1948*.
46 Medical Department, *Annual Report 1946*.
47 Medical Department, *Annual Report 1949*, p. 60.
48 Medical Department, *Annual Report 1951*, pp. 85–86.
49 Medical Department, *Annual Report 1951*.
50 Medical Department, *Annual Report 1948*, p. 52.
51 Ministry of Health, *Annual Report 1975*, p. 11.

9 SATA

The community against TB

The narrative of tuberculosis control in Singapore thus far has mainly been one of government-led interventions. But there was, in a real sense, more than one tuberculosis control programme: the other primarily was the work of the Singapore Anti-Tuberculosis Association, a non-governmental association familiar to generations of Singaporeans. From its inception immediately after the Second World War to the present day, SATA carried out a diverse body and volume of work in tuberculosis detection, outpatient treatment, public education, and rehabilitation. Its endeavours paralleled and at times even exceeded the government's efforts, extending the latter's reach in the community.

While it was modelled after national anti-tuberculosis organisations in the US and Britain, SATA's relationship with the Singapore state was unique. Shaped by national prerogatives and local circumstances, it highlighted both the possibilities and limits of a non-governmental health programme in post-war Singapore. On the whole, the official and autonomous anti-tuberculosis endeavours operated in tandem, although there were instances of disagreement over specific issues, particularly with the colonial government in the 1950s. In the postcolonial period, SATA's chief role was that of a service provider, focusing on matters that were its forte and expertise, namely, mass radiography and tuberculosis-related propaganda.

Laymen and experts

In June 1947, amid rising anger at the original Medical Plan sans a strong tuberculosis programme, a group of professionals, businessmen and doctors, both Asian and European, met for the purpose of forming an 'association for the prevention of tuberculosis'.[1] Many of those present at the meeting had been prisoners of war during the Japanese occupation. Interned at Changi Prison and Sime Relocation Camp, they had ruminated about what they could do for Singapore after the war.[2] One of the suggestions was to tackle the tuberculosis problem, which was not surprising because it was of growing concern before the war and because in their midst were a number of doctors concerned about the disease. Upon their release after the Japanese surrender, the internees formed a discussion group, which decided to establish an anti-tuberculosis association. This was approved by the government, and though it would not provide funding for the association, the connection between the two entities was evident from the start.[3]

Still, SATA was a non-governmental organisation and not merely an appendage of the colonial state. Among the 42 attendees at the 1947 meeting were S.H. Peek, the President of the Singapore Rotary Club; J.L. Wilson, Lord Bishop of Singapore; Lee Kong Chian, a respected businessman and owner of Lee Rubber; Bishop Edwin F. Lee, head of the Methodist Church of Malaya; and W.H. Jowit, manager of the Mercantile Bank of India. In his speech, Peek emphasised the importance and urgency of tuberculosis prevention, given the 'lack of foresight, administrative errors, apathy and ignorance which have accumulated over many generations'. He pointedly referenced Vickers' claim that the government had no funds to tackle tuberculosis, describing the proposed association as 'a work in which the layman and the expert can unite ... [and] contribute according to our means and qualifications'.[4]

Another person present at the meeting was Dr Chen Su Lan, the President of the Alumni Association of the King Edward VII College of Medicine, which had criticised the official disinterest in tuberculosis manifest in the original Medical Plan. Chen advocated the need for action now rather than mere 'noise'.[5] Also in attendance were Dr B.R. Sreenivasan, a tuberculosis specialist who was formerly the Chief Medical Officer of TTSH and another member of the Alumni Association, and Dr Benjamin Chew, who had resigned from Tan Tock Seng Hospital in 1946 over the government's lack of concern about tuberculosis. These individuals were the 'public-spirited people' who sought to organise a more positive response to tuberculosis compared to the British military administration, which briefly governed Singapore from after the war until 1948.[6]

The meeting concluded by unanimously establishing the Singapore Anti-Tuberculosis Association. Its 14 founder-members comprised prominent businessmen, doctors and professionals, many of whom had links to the colonial administration, namely, Peek, Chen, Sreenivasan, Lee Kong Chian, Tan Chin Tuan, Lien Ying Chow, Syed A.M. Alsagoff, R. Jumabhoy, N.A. Mallal, Sandy G. Pillay, L. Cresson, Dr P.S. Hunter, Dr B.R. Johns, and Dr L.E.C. Davis.[7] Peek, who had chaired the discussion group of ex-internees, became the first Chairman of SATA and the association's medical planning committee comprised the doctors Chen, Sreenivasan, Chew, Davis, Johns, and Hunter. SATA's Patron and President were Malcolm MacDonald, the Governor-General of Malaya, and Franklin Gimson, the Governor of Singapore, respectively – another sign of the official connection. The association's logo was the double-barred Lorraine cross and shield.

The service provider

Other than the Rotary Tuberculosis Clinic in TTSH, SATA was the main non-governmental association to devote its work to tuberculosis. It worked closely with the colonial and postcolonial administrations, particularly the Medical and Social Welfare Departments. Besides the modest funding it received from the state, SATA's work was supported mostly by public subscriptions and fund-raising (notably through the sale of its Christmas greeting seals, which started in 1950) and to a small extent by patient fees; SATA's services, unlike those of the government and TTSH, were not free. The association struggled constantly with funding throughout its history, which limited its work to X-ray screening, propaganda and rehabilitation, although it also carried out a small volume of out-patient treatment, home visits and case-finding. SATA's tuberculosis patients who required in-hospital treatment were referred to TTSH, including those unable to pay for their treatment and a small number who had their pre- and post-surgical treatment at the hospital.

SATA thus stood apart from TTSH and the government in these ways. Some patients were unhappy with the association's fee-paying policy and demanded free treatment, which the government offered.[8] SATA also ran a home treatment service for seriously ill and infectious patients. This was criticised by some doctors for being wasteful in prescribing antibiotics and collapse therapy to patients with advanced or extensive tuberculosis, which again differed from the government's stance. SATA defended its position, arguing that in its experience, patients' recovery depended as much on the individual's innate constitution – and the resistance of the Chinese was believed to be particularly high – as on the progress of the disease.[9] This commitment to treat advanced cases also led SATA to assume the difficult and unwanted work of rehabilitating chronic and unemployable patients, as we will see later in this chapter.

Despite the financial constraints, SATA's work quickly exceeded the Rotary clinic's outpatient programme, becoming one of the leading non-governmental organisations in Singapore concerned with healthcare. In 1949, SATA started a temporary clinic for diagnosis and outpatient treatment, operating out of a wooden hut next to St. Andrew's Mission Hospital in Tanjong Pagar, which in the previous year had been carried out within the hospital premises. The clinic helped to ameliorate the shortage of bed space at TTSH in a small way, but it soon became inadequate for the growing volume of work.

In 1952, with public donations and government funding, SATA built a bigger clinic, named the Royal Singapore Tuberculosis Clinic, at Hoe Chiang Road in Shenton Way in the town area. The clinic had facilities for X-ray screening, out-patient treatment, clinical tests, and rehabilitation. Five years later, it was renamed the Royal Singapore Chest Clinic to avoid (not very successfully) the stigma associated with the disease.[10] In 1955, SATA established a mobile treatment unit to operate in the rural area – the first of many vehicles which were subsequently nicknamed the 'SATA buses' for their work in the outlying areas of Singapore.[11]

While much of the ensuing work was shouldered by the state, SATA kick-started the BCG vaccination programme in 1950 following Johannes Holm's visit to Singapore. It used an experimental dry vaccine, using a preparation from the Pasteur Institute in Paris. S.H. Garlick, a radiologist and former colonial officer who became SATA's first Medical Director, was a strong proponent of the BCG vaccine. In addition, the Royal Singapore Tuberculosis Clinic's almoner and health sisters carried out home visits and case-finding work similar to their counterparts at the Medical Department. At the end of 1952, the clinic was treating 2,700 tuberculosis inpatients and 138,872 outpatients (some of whom were on the Tuberculosis Treatment Allowance Scheme) and performing 50,924 X-ray screenings. These numbers climbed steadily into the late 1950s: in 1958, the clinic treated 2,412 new patients and 222,143 outpatients and conducted 110,041 X-rays. In 1959, home visits were extended to the mothers' 'resting month' following the birth of the child, in which period they would usually not visit the clinic for treatment.

In 1953, SATA started what would become a landmark innovation – a tuberculosis insurance scheme – enabling the disease to be detected at an early stage through regular X-ray checks. The scheme also had the added benefit of producing a steady stream of revenue for the association. Insurance schemes were a distant proposition for low-income families struggling to eke out an existence, but post-war Singapore was moving economically towards the regular employment which would make insurance feasible. By 1959, more than 17,000 people were covered by the scheme, of which 73 – or more than four per 1,000 persons – subsequently developed active tuberculosis. This was a high rate of infection, underlining the need for early and routine screening.[12]

Although SATA was initially formed in response to the lack of a firm anti-tuberculosis policy by the government, its diagnostic and outpatient work paralleled the official tuberculosis control programme. As discussed elsewhere in this book, SATA occasionally disagreed with the government over substantive issues: namely, Singapore's need for a second tuberculosis clinic, the creation of a centralised case registry, and the PAP's push for compulsory BCG vaccination. Tellingly, the last two issues contained the element of compulsion to which SATA objected. Nonetheless, there were stronger strands of collaboration alongside these differences, such as SATA's support for the government's outpatient treatment, X-ray screening and case-finding programmes, while the Royal Singapore Chest Clinic was represented in the Tuberculosis Treatment Allowance Advisory Committee.

The establishment of the Tuberculosis Control Unit had initially cast doubt over the role of SATA, especially as the association had objected to compulsory notification. TBCU centralised the country's anti-tuberculosis work in the government, allowing it to govern and transform the lives of patients and their families much more than in the past. The colonial government tried to allay SATA's concerns over its future. Speaking on the proposed TBCU in 1956, the Minister for Health A.J. Braga described it as 'the first time the Government itself is attacking tuberculosis in a purposeful manner', while expressing hope that the disease would soon, like malaria, be eradicated. He assured that although TBCU would also carry out X-ray screening, SATA nonetheless remained relevant in tuberculosis control.[13]

This was largely what transpired and continued into the postcolonial period. A year after the government's case-finding survey in 1958, SATA launched a similar voluntary case-finding project, which would screen about 20,000 people above the age of 13 in the inner city district of Tanjong Pagar.[14] The association framed this survey in military terms: 'For the first time in 12 years, therefore, we have assumed the strategy of attack rather than defense'.[15] Tanjong Pagar, interestingly, was the constituency of the new prime minister of Singapore, Lee Kuan Yew. The government's Department of Information Services provided SATA with a van to announce the survey to the residents during and after the event.[16] The survey found nearly 5 per cent of the screened population to have tuberculosis. This was alarming to SATA as its clinic had operated in the district since 1952.

The finding reinforced the argument for a mass X-ray sweep of the general population, in contrast to waiting for patients to come to the clinic. The pilot screening at Tanjong Pagar turned into the truly offensive campaign, which was the PAP government's mass X-ray survey in the 1960s. SATA concentrated on rural and urban districts such as Delta and Pasir Panjang while also lending its mobile X-ray units to the government's surveys. Tanjong Pagar was surveyed a second time in 1963 by the association to determine the epidemiological situation in the district, with only 1.4 per cent of the people surveyed found to have active tuberculosis this time. Similarly, SATA's resurveys of two rural districts – Siglap and Kampong Kembangan – within the space of two years (1963 then 1965) showed a reduction of cases by two-thirds.

In the 1960s, SATA cemented its role as a provider of anti-tuberculosis services in the community. Its relationship with the state grew even closer as the PAP government expanded its tuberculosis control programme. Membership of the association's tuberculosis insurance scheme rose, with more firms realising the value of a tuberculosis-free staff. More to the point, the growth of manufacturing and commerce, based on foreign and state capital investment, was creating full-time jobs in Singapore, which included the provision of medical benefits. In 1964, there were more than 37,000 people on the scheme, and X-rays done at pre-insurance screenings uncovered 63 cases of active tuberculosis, while 103 insured persons developed the disease during the year and received treatment. In 1969, of the 125 persons granted sick leave for their treatment, the average length of time away from work was only six days per patient.[17]

The typical insured patient with tuberculosis was unlikely to be a burden to his or her family or a threat to the community. In a study of 73 active tuberculosis patients in 1959, SATA found that the majority did not have to stop working, while those who had to take sick leave were able to return to their jobs within two months.[18] Early detection through regular X-rays, SATA noted in 1973, had the great benefit of unearthing the disease before it became infectious.[19] The incidence of active tuberculosis among the insured group fell from 0.42 per cent in 1959 to 0.06 per cent in 1979. By this time, however, as the association warned in 1982, 'It is ironical that the success thus achieved is beginning to jeopardize its future': some companies had begun to rethink their subscription to the insurance scheme. SATA warned that tuberculosis epidemics in the US had broken out within air-conditioned offices.[20]

In 1962, SATA built a regional diagnostic and outpatient clinic, the Uttamram Chest Clinic at Upper Changi Road, which served the eastern part of Singapore.

The clinic was named after G. Uttamram, a businessman and SATA council member who had donated the land to the association in 1954 (it was later renamed the SATA Uttamram Clinic). X-rays continued to drive SATA's work in the community, their number rising from 110,041 in 1958 to 154,821 in 1963. During the period from 1950 to 1979, SATA performed X-rays on more than 2.6 million people, of whom 54,040 (or 0.02 per cent) were found to have radiologically active tuberculosis.

In addition, financial assistance and free food rations, sometimes provided as part of the government's tuberculosis treatment allowance scheme, continued to be given to the small and declining numbers of SATA's low-income, usually older patients. In 1969, an average of 231 patients and 312 dependants received such help each month. SATA also continued to organise home visits to patients who were too ill or old to attend the clinic, and to defaulters; in 1964, over 70 per cent of the latter group resumed their treatment, largely because of these home visits. Dr N.C. Sen Gupta, its long-standing Medical Director from 1958 to 1987, took the view that doctors, and the medical system more generally, were equally at fault when patients defaulted on their treatment. The failure to complete treatment, he noted, was often due to difficulties on the patients' part: those who were still in employment may not be able to find the time or means to travel to a clinic, or did not do so for fear of being dismissed if their employer found out that they had tuberculosis.[21]

As with the government's health policy and developments at TTSH, SATA extended its work into heart and non-tuberculous chest diseases in the mid-1960s as the disease declined as a national threat. In 1965, the association set up a new department of cardiology. Two years later, its tuberculosis clinic was renamed the SATA Chest and Heart Clinic. In 1972, SATA finally ceased its mass X-ray survey due to low attendances (less than a quarter of the eligible population in some districts) and yields. These were problems it had already encountered as early as 1964, partly due to growing fears of the hazards of radiation from the X-rays.[22] In 1986, SATA organised the 26th International Union Against Tuberculosis World Conference on Tuberculosis and Respiratory Diseases in Singapore, which institutionalised the growing global focus on chronic non-tuberculous chest diseases.

From the 1970s to the 1990s, tuberculosis continued to be part of SATA's remit, albeit in lessening importance as a growing proportion of tuberculosis notifications were made to TBCU. In the late 1990s, SATA provided free X-ray screening to elderly persons at community centres, in line with the government's policy, while about a quarter of its tuberculosis patients who needed assistance received free treatment. SATA duly supported the government's Singapore Tuberculosis Elimination Programme launched in 1997, the 50th anniversary of the association's founding. On the existing threat of tuberculosis, SATA warned that 'There is now a dangerous level of complacency about the disease'.[23] For a long time, it had publicly emphasised the need to reduce the incidence of tuberculosis to a single digit per 100,000 persons, as in many other developed countries.[24] This brings us to SATA's public education programme.

Other battlefronts

SATA's consistent view of the education of the public on tuberculosis was that 'Propaganda is actually a main battle-front activity. It is also the fundamental purpose of SATA'.[25] It envisaged health education in a broad way, based on the belief that 'by far the most effective of all forms of propaganda is the treatment of a patient', beginning with the patient and his or her family at home and thus critically connected to the work of the almoner and health visitors.[26] In the early 1950s, both the public and domiciliary wings of SATA's propaganda were effective in persuading sufferers living in the community that the association would provide medical and financial assistance to them, even if its services were not free.[27] In 1954, SATA extended its public education to schools, with nearly 3,500 students and teachers visiting its premises at Shenton Way to learn about tuberculosis, while others did so within the schools. The propaganda addressed pertinent issues such as the nature of tuberculosis and its treatment, the implications of defaulting from treatment and the social stigma.

These efforts continued after 1959. The new SATA Chairman, journalist Wee Kim Wee, launched a bimonthly periodical, *SATA Bulletin*, which carried news on the disease in the four major languages of Singapore (the Tamil issue commenced in 1963). The bulletin was distributed to schools, community centres, businesses, clinics, and the public. In 1960, SATA and the government jointly produced a film and three radio programmes on tuberculosis.[28] SATA's activities were also broadcast on Rediffusion, a cable radio service. The association continued to work with the Ministry of Health to organise lectures, film screenings and exhibitions on tuberculosis for secondary school children and the public. In 1980, SATA stressed that 'keeping the word "tuberculosis" before the public is a contribution towards guarding the health of the community'.[29] In 1982, the newly built Information Centre, at its new office in Cantonment Road, continued to disseminate information to the public on the control and treatment of tuberculosis and other diseases.

The focus of SATA's work in the postcolonial period was mass radiography and early detection. These measures comprised a sizeable component of its public education efforts; they also highlighted a similar reflexivity towards tuberculosis control as the government's. Concurring with the latter, SATA cautioned in 1973 that although the incidence of the disease had fallen, there was still a hidden pool of infection in the community, which may become active.[30] In 1979, the association pointed out that only 7.3 per cent of the 215,860 people it X-rayed came forward of their own accord (likely because they had a persistent cough). The majority of the screening had been mandatory under the TB and Heart Insurance Scheme, which had become the mainstay of SATA's case-finding programme. Most people were no longer voluntarily coming for X-rays, and its screening efforts, the association urged, should be further intensified.[31] But the push for mass radiography ceased in 1981 when the government accepted the health risks posed by radiation.[32]

In the 1970s and 1980s, SATA remained concerned with the relatively high incidence of tuberculosis in Singapore, lamenting that 'Public apathy has to be blamed to a large extent for this unsatisfactory state of affairs since this is a disease which can be completely cured'.[33] As noted in Chapter 7, Dr Sen Gupta was tireless in warning against complacency, especially in the prevalence of tuberculosis among older teenagers and elderly people. The latter, SATA pointed out, were an important group because they were often caregivers of their young grandchildren, and also because some of them were in financial difficulty and unable to go for an X-ray. In 1980, SATA also underlined the need to devote greater attention to the incidence of the disease among females who, unlike males, were younger (under 30 years of age). Throughout the decade, the association beseeched the government to undertake a follow-up study to the 1975 national tuberculosis prevalence survey in order to assess and revise the existing control programme.

South winds

While the treatment of tuberculosis became increasingly effective after the war, the care and rehabilitation of two particular groups of patients – chronic sufferers and cured patients who were unemployable – became a pressing issue. Although these two groups had different medical statuses, they shared a common problem: employment, which could no longer be affected by treatment, and both were also affected by the social stigma against tuberculosis.

The lack of beds at TTSH in the immediate post-war years affected the chronic sufferers in particular. The priority for admission was based on medical grounds and thus given to cases that could improve with treatment. Most chronic sufferers were not treated as outpatients or if at all, although a small number of advanced cases were accepted 'more on a social priority' because they had been abandoned by their families or had no means to support themselves.[34] The majority of chronic cases seeking admission to TTSH were males, such that the Medical Department was unable to abide by the allocation of 68 beds to males and 32 to females in 1949. The following year, the hospital increased the male ratio, reserving 75 beds for males and 25 for females.

In 1950, the Medical Department confessed that 'The vast majority of these cases have to stay outside the hospital'.[35] This was ironic, given TTSH's origins as a pauper hospital which was now rejecting the chronic tuberculosis patients. These sufferers were likely to be physically weak, unemployable and destitute. To complicate matters, the old colonial practice of repatriating such persons had become untenable after the war as more people were born locally in Malaya and Singapore. Neither were they often able to apply to the Tuberculosis Treatment Allowance Scheme, being eligible only for the smaller allowances paid out under the Public Assistance scheme. As tuberculosis specialist R.G. Grove-White revealed in 1956, Singapore's allowance scheme followed the T266 programme in wartime Britain, where it was not conceived 'primarily as a public health thing'. The scheme instead had to produce 'dividends' by curing patients and returning them to the workforce.[36] This narrow scope placed the scheme beyond the most needy patients.

The government's almoners visited a small number of chronic patients at their homes, attempting to persuade their landlords or co-occupants not to evict them. They tried to use the government's Welfare Fund or Public Assistance fund to help pay the patients' rent, or sought help from the voluntary welfare homes run by the Hakka and Hainanese communities in Singapore to feed the chronic patients, or from the Kwong Wai Sui Free Hospital, which catered to the Cantonese.[37] But such piecemeal, community-specific efforts failed to help numerous chronic sufferers, many of whom were rejected by their neighbours and community and driven out of their homes into the five-foot-ways and streets.[38] Suicide, as a 1954 medical study highlighted, was the final course of action taken by many desperate chronic sufferers, especially men with families who were weighed down by the twin burdens of unemployment and illness.[39]

The question of chronic patients divided the government and medical community. At one extreme were doctors such as S.H. Garlick (SATA's Medical Director) and international experts Andrew Morland and Frederick Heaf, who supported the establishment of isolation camps. In 1950, Morland proposed that such camps, if built to appeal to chronic and cured patients, would play a major role in the anti-tuberculosis programme.[40] In 1952, Heaf called for a more 'dynamic' programme in Malaya, in which chronic patients would be discharged from hospitals to free up beds for those who were treatable.[41] The Medical Department also thought favourably of isolation homes funded by the community but offering little treatment, which would free up 'Many precious beds in our present hard-pressed hospitals'.[42]

These expedient views lay at the heart of the state's treatment policy, which leaned towards treatable patients, but the idea of isolation camps was contentious and never implemented. This was partly because they involved compelling chronic patients to live there, but also because they would be costly. There was the additional question, as the *Straits Times* pointed out, of who would bear the responsibility: would such a home come under the charge of the government or a voluntary association?[43] The government deemed it to be the community's responsibility. In response to Morland, some government doctors believed that the camps could work in Malaya but not in Singapore, where hospitals were 'hopelessly underbedded and understaffed', and would not be able to care for the large numbers of chronic patients.[44]

For the second group, the cured and recovering patients, the Medical Department envisaged that they would be restored to a productive place in the community. However, although certified fit for light work, many of them belonged to the working class and lacked the education or skills to find alternatives to manual work, which was not suited for their poor health. With the help of volunteer social workers, TTSH organised modest programmes of occupational and diversional therapy, allowing patients to learn elementary skills such as tailoring, sewing (the Singer Sewing Machine Company offered machines at a discounted price), basket-weaving, knitting, and toy-making. The hospital also organised a 'six months' scheme', hiring recovering patients as paid employees to perform light work in the hospital. Such programmes, though appreciated by patients in the 1950s, were merely limited, short-term measures.

But while medically cleared for work, many discharged patients had difficulty doing so due to various reasons. Without adequate assistance, several had become unemployable like the chronically ill patients. Some cured patients had stayed in the hospital for such a long duration that they had lost contact with their kin and friends, or they had few of them in the first place.[45] Stigma was again a major hurdle, and some cured patients were rejected by co-workers who knew about their illness. On the challenges facing the newly formed Rehabilitation Committee in the Ministry of Labour, S. Graham of the almoner's division observed in 1951 that many employers were unwilling to accept cured patients, though the committee deemed its efforts 'an educational experiment which will perhaps bear fruit later on'.[46] For a small group of patients, the almoners were able to find loans for them to set up or revive a business. More generally though, the problem of employment was structural rather than individual: post-war Singapore was still largely dependent on the entrepôt trade and did not possess a sizeable industry to provide enough jobs for a growing population.

The care of chronic and cured patients ultimately devolved by default to the community, even though the colonial government was much criticised for evading its duty to the people in the early 1950s.[47] The community already maintained a number of homes for ill and destitute persons, but these plainly were not sufficient. In 1951, the Government Tuberculosis Advisory Board and Chinese Advisory Board jointly endorsed the establishment of a home for chronic and destitute tuberculosis patients, which would be funded by the Chinese community. The home would initially only have 100 beds – a miniscule figure compared to ten times the number being turned away from TTSH at the time. Two years later, a Singapore Tuberculosis Home Association was formed to organise the setting up of this paupers' home.

A few individual narratives outlined in a 1954 SATA publication highlight the struggles of tuberculosis patients and the key roles played by almoners and nurses. 'T.J.L.', a former hawker, was a 'sad case' who resided in a wooden kampong house. His wife had left him alone with their ten-year-old daughter when he could no longer work because of his illness. His condition was once treatable but had become chronic due to a poor diet and long hours of continuous work. A staff nurse from the almoner's department collected on his behalf his tuberculosis treatment allowance – his only means of support – and a supply of eggs and milk. The Social Welfare Department would take custody of his daughter when he passes away. By contrast, 'T.W.L.', a young girl who was once thin and anaemic, was able to make a good recovery, aided by rest, nutritious home food and a prescription of streptomycin and isoniazid. Reportedly, she 'has a good colour, is feeling well, and is putting on weight'. Another recovering case, 'C.A.N.', received government assistance to move from a dim, eight-square-feet shophouse cubicle to a modern Singapore Improvement Trust flat. He also received twice-weekly injections and tablets, which helped him to gain 30 pounds.[48]

The rehabilitative role was eventually inherited by SATA. In 1953, it formed a rehabilitation division to assist a small but growing number of recovering and cured patients. Working with the Labour Department, the division tried to help 140 patients find employment in the trade unions, business firms and the War Department. A small tailoring centre was established within the premises of the

Shenton Way clinic that year, which later added embroidery, dress- and box-making, book-binding and printing, and carpentry. The trainees received small allowances of $50 per month while those who remained at the centre after six months of training were paid for piecework, making from $60 to $200 per month. They remained under supervision, working limited hours so as not to over-exert themselves.[49] However, a growing number of recovering patients in the 1950s were unwilling to take up such work, as it differed greatly from their previous employment, and insisted on collecting their treatment allowance until they were well.[50]

In 1954, due to the lack of space at the centre, the association built a bigger rehabilitation home located at the South Winds Hotel at Tanjong Balai in Jurong. This was, in an ironic twist of history, the rejected site of the proposed second sanatorium of Singapore. Following the rejection, Legislative Councillor Dr C.J. Palgar had suggested the hotel to be used to accommodate chronic tuberculosis cases who were unable to gain admission to TTSH. The owner of the site was rubber businessman and philanthropist Lee Kong Chian, one of the founding members of SATA. He loaned the site to the association and was willing to sanction its use for a purpose in the public interest.[51]

The new rehabilitation centre at South Winds was a small one, serving initially just 22 cured tuberculosis patients, who were taught husbandry and craftsmanship to help them become financially independent. In 1956, the number of patients and ex-patients living at the settlement rose to 40. Unable to find employment outside, they were engaged in various forms of low-paying work: chicken farming, pig rearing, vegetable- and flower-growing, and rattan furniture-making. Many of them initially had reservations towards living in the settlement, fearing the work to be heavy, but they were persuaded that it would not be so, and that light work would gainfully occupy their mind and improve their health. The settlement also offered English-language classes in the evening, although it is not known if that was an economic benefit to the patient-residents.

However, South Winds was not able to become self-supporting and remained underfunded, which limited its expansion.[52] The number of patient-residents continued to rise slowly, reaching 88 in 1959, although 20 individuals left without having completed their training. By then, six residents had brought their families to live with them in the settlement.[53] All these pointed to a limited effort to deal with the issue of rehabilitation in the 1950s, which managed to impress most visitors to the home.[54]

The problem of chronic and unemployable patients was never truly resolved: it simply went away in the postcolonial period. In 1959, the Medical Department emphasised that 'The old problem of employment and re-habilitation after discharge are [sic] more pressing than ever'.[55] The following year, TTSH accepted a small number of chronic infectious ambulant cases who were previously warded at Woodbridge Hospital. The South Winds site was acquired by the PAP government sometime in the early 1960s for the development of Jurong Industrial Estate – its flagship industrialisation project. The rehabilitation of chronic and unemployable patients was subsequently limited to the rehabilitation workshop attached to the Royal Singapore Chest Clinic in Shenton Way.[56]

In 1963, SATA pronounced that the need for rehabilitation had 'diminished considerably in recent years'.[57] The following year brought an end to the weaving and rattan furniture work, with the skills confined to tailoring, sewing and printing. The budget for rehabilitation was reduced, although an effort was made for the patients to make simple products for tourists to Singapore. Dr Sen Gupta pointed out that rehabilitation diminished as a problem, partly because chemotherapy was effective in most cases, allowing patients to return fully to work, and partly because the social stigma which had previously made them unemployable had faded away.[58] SATA's rehabilitation work ceased in 1980 when its Shenton Way site was acquired by the government and it moved to much smaller premises, a third of its former size in Cantonment Road.

The history of the chronic and unemployable patients, usually older persons, is mostly untold. The rare story of Poon Siew, a painter of buildings, printed in a newspaper article in 1974, illustrates the plight of this small group of chronic tuberculosis sufferers. Around 1954, he contracted the disease while working overseas in Brunei and was treated for it. He returned to Singapore and was diagnosed with advanced tuberculosis five years later, at the age of 52. He was warded in TTSH and received both injections and pills, but with little improvement to his condition, he was taken off treatment in 1964. Poon's story illuminates his difficulties in following the treatment regimen while continuing to work as a painter to support his wife, a homemaker who eventually left him, and eight young children:

> The doctors advised me to rest and eat well. But how could I? A painter's job is hard, climbing up and down scaffolding and ladders all day long. I just couldn't take things easy.... My income was not much and good food was too expensive.... Worried? What's the use of worrying? It was just one of those things that happened. It depends on your outlook in life.[59]

In 1979, reflecting on the past, SATA surmised that few anti-tuberculosis organisations in the world could match its achievements in combating the disease.[60] A decade later, the association was renamed SATA CommHealth, with its main portfolio to promote affordable community health in the country. Having made its mark in the provision of services for the surveillance, detection, treatment, and public education of tuberculosis, it would now do the same for heart disease and community health. Remarkably, SATA's sustained role and work had bridged the colonial and postcolonial periods. Tuberculosis control had not only defined the shape of the Singapore state, but also that of the country's successful anti-tuberculosis association.

Notes

1 *Malaya Tribune*, 4 June 1947.
2 SATA, *The Royal Singapore Tuberculosis Clinic of the Singapore Anti-Tuberculosis Association* (Singapore: D. Moore, 1954); *Singapore Free Press*, 27 September 1961.
3 Lim Kay Tong and Mary Lee, *Fighting TB: The SATA Story (1947–1997)* (Singapore: Singapore Anti-Tuberculosis Association, 1997).

4 *Malaya Tribune*, 4 June 1947.
5 *Malaya Tribune*, 4 June 1947.
6 Oral History Centre, National Archives of Singapore, Interview with Chew Chin Hin, Reel 1, 7 July 1999; Interview with N.C. Sen Gupta, 5 February 1999.
7 Lim and Lee, *Fighting TB*.
8 'SATA and the Un-Co-operative Patient', in SATA, *The Royal Singapore Tuberculosis Clinic of the Singapore Anti-Tuberculosis Association* (Singapore: D. Moore, 1954), pp. 95–97.
9 SATA, *The Royal Singapore Tuberculosis Clinic*, p. 12.
10 *Singapore Free Press*, 14 September 1957.
11 SATA CommHealth, *The SATA Story: Celebrating 65 Years of Caring for the Community* (Singapore: SATA CommHealth, 2012), p. 8.
12 *Straits Times*, 15 March 1957.
13 *Straits Times*, 6 October 1956.
14 DIS 161/58, Singapore Anti-Tuberculosis Association Press Release, 30 June 1959.
15 SATA, *Annual Report 1959*, p. 2.
16 DIS 161/58, Memo George G. Thomson to Dr N.C. Sen Gupta, 18 June 1959.
17 SATA, *Annual Report 1969*.
18 SATA, *Annual Report 1959*.
19 SATA, *Annual Report 1973*.
20 SATA, *Annual Report 1982*, p. 6.
21 Oral History Centre, National Archives of Singapore, Interview with Sen Gupta, Reel 3, 5 February 1999.
22 Lim and Lee, *Fighting TB*; SATA, *Annual Report 1964*.
23 Lim and Lee, *Fighting TB*, p. 15.
24 SATA, *Annual Report 1988*.
25 SATA, *The Royal Singapore Tuberculosis Clinic*, p. 12.
26 SATA, *The Royal Singapore Tuberculosis Clinic*, p. 35.
27 Medical Department, *Annual Report 1953*.
28 SATA CommHealth, *The SATA Story*.
29 SATA, *Annual Report 1980*, p. 14.
30 SATA, *Annual Report 1973*.
31 SATA, *Annual Report 1980*.
32 *Straits Times*, 19 September 1981.
33 SATA, *Annual Report 1984*, p. 6.
34 Medical Department, *Annual Report 1949*, p. 129.
35 Medical Department, *Annual Report 1950*, p. 129.
36 Ivy L. Foo, 'Tuberculosis Health Visiting in Singapore', Pan-Malayan Tuberculosis Conference, *Transactions of the First Pan-Malayan Tuberculosis Conference 1–4 November 1956* (Singapore: Government Printing Press, 1957), p. 5.
37 *Straits Times*, 30 October 1950.
38 O.B. Leathart, 'The Almoner's Work in Connection with Tuberculosis in Singapore', *Transactions of the First Pan-Malayan Tuberculosis Conference*.
39 *Straits Times*, 22 November 1954.
40 'Report of Dr. A. Morland on Tuberculosis in Malaya', *The Medical Journal of Malaya* 4 (4), June 1950.
41 *Straits Times*, 13 May 1954.
42 Medical Department, *Annual Report 1952*, p. 4.
43 *Straits Times*, 13 May 1954.
44 'Report of Dr. A. Morland on Tuberculosis in Malaya', p. 280.
45 Leathart, 'The Almoner's Work in Connection with Tuberculosis in Singapore'.
46 Medical Department, *Annual Report 1951*, p. 136.
47 *Singapore Free Press*, 26 May 1952.

48 Pearl Stewart, 'The Visiting Nurse Speaks', in SATA, *The Royal Singapore Tuberculosis Clinic*, pp. 88, 89.
49 B. Jensen, 'Rehabilitation of Tuberculous Patients', *Transactions of the First Pan-Malayan Tuberculosis Conference*.
50 'SATA and the Un-Co-operative Patient', in SATA, *The Royal Singapore Tuberculosis Clinic*, p. 96.
51 *Straits Times*, 10 March 1953.
52 Jensen, 'Rehabilitation of Tuberculous Patients'.
53 SATA, *Annual Report 1959*.
54 *Singapore Free Press*, 14 September 1957.
55 Ministry of Health, *Annual Report 1959*, p. 159.
56 *Straits Times*, 9 July 1960.
57 SATA, *Annual Report 1963*, p. 9.
58 Oral History Centre, National Archives of Singapore, Interview with Sen Gupta, Reel 5, 5 February 1999.
59 *New Nation*, 29 October 1974.
60 SATA, *Annual Report 1979*.

Conclusion

Writing the long history of tuberculosis in Singapore is a complex undertaking. While to a large extent our book charts a narrative of activity and accomplishment, we have also uncovered new perspectives, nuances and limitations in the anti-tuberculosis programme. The history is tied to Singapore's early development as a British colony and its transition to a nation-state. It highlights connections between tuberculosis and urban history, from the first surveys of the shophouses at the start of the twentieth century to the visits of almoners and doctors to patients' homes since the 1950s and 1960s. Furthermore, as our book discusses, the open city-state of Singapore is closely linked to the outside world. This is evident in the international systems of medical expertise and collaboration, beginning with W.J.R. Simpson's 1907 study of sanitation in the town, but also in the immigrants, both historical and current, who carry the disease.

The houses and migrants of Singapore take the narrative of tuberculosis control beyond hospitals and doctors to the urban history of the city-state. The Singapore state presided over the control programme, shaping the medical lives and social behaviours of patients and their families. It also largely defined the role of the community, specifically the work of the Singapore Anti-Tuberculosis Association as a medical service provider. The state gained much confidence and prowess in combating tuberculosis over a lengthy period of time, due largely to the political commitment of the People's Action Party government in the 1960s, but also to the foundational work laid by the colonial regime and medical doctors even before the Second World War. Conversely, the state adopted a reflexive attitude towards the illness from the 1970s onwards, which shaped the PAP's governance of Singapore society as a whole. Both government officials and tuberculosis physicians have remained concerned about the incidence of tuberculosis in the community in recent years, particularly among the elderly. As Dr N.C. Sen Gupta of SATA emphasised in an oral history interview in 1999, despite its growing emphasis on heart and chest diseases, the association should remain focused on tuberculosis.[1] Such reflexivity is built into the national psyche; a newspaper article in 2005 called for 'constant vigilance' against an apparent increase in the disease among school students and staff.[2]

Nevertheless, Singapore's relative success in tuberculosis control has not been absolute. The study of history brings to the fore issues that were not salient even to the state. One of them, manifest in the inter-war period, was the tendency to view the disease in racial terms, which persisted even after the theory of racial immunity and resistance had been discredited. Racial classification of tuberculosis remains part of the system of governance in Singapore and was rooted in colonial terminology and practice. But the utility of the concept of race was suspect: the illness affected particularly low-income workers residing in congested housing, suffering from poor health or coming from countries with a high prevalence of the disease.

In Singapore, there were other historical shortcomings, failings, and blind spots to which the administration should have been more alert: patients and contacts avoiding screening or not completing their treatment, a significant minority of drug-resistant cases, social stigma against tuberculosis among employers and the public, and migrant workers with tuberculosis. On the latter, as Dr Chew Chin Hin recalled in an oral history interview, Singapore had initially not demanded X-ray screening of migrant workers from Asian countries who may have arrived with an infection, or even a drug-resistant infection.[3] Many of the old problems have persisted to the present day.

Despite the achievements of the last 50 years, tuberculosis remained a significant problem in Singapore. The incidence of the disease among Singapore residents fell to a low of 35.0 per 100,000 population in 2007, only to subsequently rise and hover around 40.0 per 100,000 population since,[4] with 1,586 new cases notified in 2018. In a recent study commissioned by the Ministry of Health in 2014, 1,690 randomly selected adult Singapore residents were tested for latent tuberculosis.[5] The overall prevalence of latent tuberculosis in this sample was 12.7 per cent, increasing from 2.4 per cent in those aged from 18 to 29 years to 23.2 per cent in those aged 70 to 79 years. Male subjects were more likely to test positive than females, regardless of age, whereas those born in Singapore were less likely to test positive compared to older persons born overseas, with the highest prevalence present in those born in India.[5]

The stagnation of Singapore's incidence of tuberculosis stands in contrast to what has been occurring in most countries worldwide, with the global incidence rate of tuberculosis falling at an average of 2 per cent each year.[6] The major factors for this stagnation have remained unchanged, including an ageing local population and an expanding migrant worker population drawn from neighbouring countries with a high incidence of tuberculosis.[7] The median age of the population has increased from 19.5 years in 1970 to 40.8 years in 2018 with the old-age support ratio (the number of residents aged 20 to 64 for each resident aged 65 and older) falling from 13.5 to 4.8 in the same period, while the non-resident population – which includes migrant workers – surged from 60,900 in 1970 to 1,644,400 in 2018.[8]

Drug resistance, an old issue, has become a major global problem in tuberculosis control. Rates of multidrug-resistant tuberculosis (MDR-TB, where the organism is resistant to the key anti-tuberculosis drugs) have crept up worldwide, with over half a million persons newly infected in 2017.[6] The situation in Singapore was more nuanced, with under 2 per cent of all new cases of TB being MDR-TB since 2000, of which the majority are believed to be imported into Singapore as opposed to local transmission.[9] But this changed in 2012, when a cluster of MDR-TB cases was traced to two local area network (LAN) gaming centres, which triggered off a secondary cluster of cases in a block of public housing flats. In both situations, the Ministry of Health took rapid and extraordinary steps to contain the outbreak, including offering onsite screening for the entire block with the use of a mobile X-ray bus – a throw-back to the past.[10] Singapore is located in a region with a high burden of MDR-TB.[6] The threat of importation of MDR-TB and its uncontained local spread is significant, because drug-resistant tuberculosis requires longer courses of therapy with drugs that have greater risks of adverse effects and are also more costly.

Social stigma against tuberculosis is another historical issue: it remains present in Singapore despite decades of public health education. This may deter sufferers from seeking help, or affect their livelihood even after treatment. Seah See Seng – one of the cases of MDR-TB at the residential block mentioned above – recounted in an interview that his meagre earnings from selling tissue fell and some of his neighbours shunned him after they found out about his illness.[11] Singapore has legislated against the socio-economic consequences of such stigma, primarily in the area of employment. Under the Employment and Industrial Relations Act, the Minister for Manpower Lim Swee Say told Parliament in 2016 that 'employees and union members in unionised companies … can appeal to MOM [the Ministry of Manpower] for reinstatement if they have been unfairly dismissed'.[12] He also urged employers to treat these employees 'with fairness and compassion' in order to minimise the risk of transmission of tuberculosis at the workplace.[12] Nonetheless, much more needs to be done to educate the public on tuberculosis, precisely because it is widely regarded as an illness of the past.

So, what does the future hold? Although still considerably underfunded, global spending on tuberculosis research has doubled since 2005, reaching approximately US$772 million in 2017.[13] This has led to the development and testing of new anti-tuberculosis drugs which, together with breakthroughs in other fields such as immunotherapy, offer the distinct possibility of shorter course, and potentially less toxic, drug regimens for both active (MDR-TB and drug-susceptible) and latent tuberculosis in the near future. New and more effective vaccines may also eventually replace the venerable Bacille-Calmette-Guérin vaccine in use since 1921. Similarly, advances in tuberculosis diagnostics offer the hope of earlier detection of tuberculosis, equally important in preventing the further spread of the disease. As a globally connected and forward-looking state, Singapore will likely tap into and benefit from these developments.

Tuberculosis has, however, always been a social and political disease in addition to a medical one, and it is in this area that Singapore may have to undertake more substantive reforms. There is some need, as Christian McMillen put it in his global survey of the illness, for tuberculosis to be 'discovered' and recognised as a major disease once again in the city-state.[14] Although it played an important role in Singapore's development as a colonial port city and subsequently as a nation-state, the disease's importance diminished alongside the incidence of the disease. Tuberculosis has been at risk of being viewed as an annoyance to policymakers and neglected relative to other major diseases. This would run counter to the history of reflexivity displayed by officials and doctors over the past century. Having a new and ambitious national target for tuberculosis control, as has been achieved by Japan, might help to focus or generate resources. The importance of socio-economic improvements, both for the general population and low-income migrant workers, and tackling the issue of stigma should also not be underestimated.

Just as crucially, new public–private partnerships have emerged worldwide, each of which have attempted to target different aspects of tuberculosis control that are difficult to achieve independently. The notion of such partnerships is salient in Singapore – historically exemplified in the collaboration between SATA and the state since the 1950s and 1960s. An updated and rejuvenated form of this relationship may involve a reorientation towards tuberculosis control on the part of SATA, and for the association to expand its function beyond that of a service provider. This may entail, more fundamentally, a reappraisal and adjustment of the role of the state which has expanded since that time in order to accommodate new stakeholders and partners. These are by no means easy changes in the local context, but they may prove decisive in achieving Singapore's part in the World Health Organization's goal of ending the global tuberculosis epidemic by 2035.[15]

Notes

1 Oral History Centre, National Archives of Singapore, Interview with N.C. Sen Gupta, Reel 6, 5 February 1999.
2 *Today*, 18 July 2005.
3 Oral History Centre, National Archives of Singapore, Interview with Chew Chin Hin, Reel 6, 7 July 1999.
4 Ministry of Health, 'Communicable Diseases Surveillance in 2016', www.moh.gov.sg/resources-statistics/reports/communicable-diseases-surveillance-in-singapore-2016.
5 P. Yap *et al.*, 'Prevalence and Risk Factors Associated with Latent Tuberculosis in Singapore: A Cross-sectional Survey', *International Journal of Infectious Diseases* 72, July 2018, pp. 55–62.
6 World Health Organization, *Global Tuberculosis Report 2018*, www.who.int/tb/publications/global_report/en/.
7 Das, S. *et al.*, 'Spatial Dynamics of TB within a Highly Urbanised Asian Metropolis using Point Patterns'. *Scientific Reports* 7 (36), 2017, pp. 1–9.
8 Department of Statistics, *Population Trends 2018*, www.singstat.gov.sg/-/media/files/publications/population/population2018.pdf.
9 C.B. Chee *et al.*, 'The Imminent Threat of Multidrug-resistant Tuberculosis in Singapore', *Singapore Medical Journal* 53, 2012, pp. 238–240.
10 Z.J.M. Ho *et al.* 'Investigating a Cluster of Multi-drug Resistant Tuberculosis in a High-rise Apartment Block in Singapore', *International Journal of Infectious Diseases*, 67, 2018, pp. 46–51.
11 *The New Paper*, 16 July 2016, www.tnp.sg/news/singapore/man-alienated-community-after-tb-treatment.
12 Written answer by Mr Lim Swee Say, Minister for Manpower, to Parliamentary Question on Unjust Termination of Employment for Employees Diagnosed with

Tuberculosis, 2016. www.mom.gov.sg/newsroom/parliament-questions-and-replies/2016/0913-written-answer-by-mr-lim-swee-say-pq-on-unjust-termination-of-employment-for-employees-diagnosed-with-tuberculosis.

13 Treatment Action Group, *Tuberculosis Research Funding Trends 2015–2017*, 2018. www.treatmentactiongroup.org/sites/default/files/tb_funding_2018_final.pdf.
14 Christian W. McMillen, *Discovering Tuberculosis: A Global History, 1900 to the Present* (New Haven, CT and London: Yale University Press, 2015).
15 World Health Organization, *The End TB Strategy*, www.who.int/tb/strategy/end-tb/en/.

Bibliography

Archival sources

Britain

CO 273 Straits Settlements Original Correspondences.
CO 275 Straits Settlements Sessional Papers, 1855–1940.

Singapore

Department of Information Services (DIS).
Director of Medical Services (DMS).
Ministry of Health (MOH).
Singapore Improvement Trust (SIT).
Social Welfare Department (SWD).
Tan Jiak Kim.
Treasury (CSO TRY).

Published sources

'A Campaign against Tuberculosis', *Medical Journal of Malaya* Vol. IV, 1930, pp. 73–74.
Chee, Cynthia Bin-Eng and Yee Tang Wang. 'TB Control in Singapore: Where Do We Go from Here?', *Singapore Medical Journal* 53 (4), 2012, pp. 236–238.
Chen, Su Lan. 'Opium and Tuberculosis', *Medical Journal of Malaya* Vol. VII, 1932, pp. 25–31.
Chew, C.H. and P.Y. Hu. 'BCG Programme in the Republic of Singapore', *Singapore Medical Journal* 15 (4), December 1974, pp. 241–245.
Choe, Alan F.C. 'Urban Renewal', in Ooi Jin-Bee and Chiang Hai Ding (eds.), *Modern Singapore* (Singapore: University of Singapore, 1969), pp. 161–170.
City Council. *Annual Reports.*
Clarke, J. Tertius. 'Tuberculosis in the Tropics', *Medical Journal of Malaya* 2 (1) 1927, pp. 19–23.
Department of Statistics. *Population Trends 2018*, www.singstat.gov.sg/-/media/files/publications/population/population2018.pdf.
Department of Tuberculosis Control. *Annual Reports.*
Education Department. *Annual Reports.*
Epidemiological News Bulletin.

Faris, D.W.G. 'Some Figures Relating to Tuberculosis in the Straits Settlements', *Journal of the Malayan Branch of the British Medical Association* 1 (3), December 1937, pp. 211–216.

Galloway, David J. 'Observations on the Death Rate', *Journal of the Malayan Branch of the British Medical Association* January 1907, pp. 1–4.

Galloway, David J. 'Notes on Some Local Aspects of Tuberculosis', *Medical Journal of Malaya* 3 (1), 1928, pp. 1–6.

Goh, Kee Tai. *Epidemiological Surveillance of Communicable Diseases in Singapore* (Tokyo: Southeast Asian Medical Information Center, 1983).

Goh, Keng Swee. *Urban Incomes and Housing: A Report on the Social Survey of Singapore, 1953–54* (Singapore: Department of Social Welfare, 1956).

Grove-White, Mary L. 'Review of the First 500 Cases in Receipt of Financial Assistance under the Tuberculosis Treatment Allowance Scheme of the Colony of Singapore', *Medical Journal of Malaya* 7 (4), June 1953, pp. 278–285.

Hanam, E. 'Heart Disease from a Case-Finding Tuberculosis Survey in Singapore', *Singapore Medical Journal* 2 (1), March 1961, pp. 3–5.

Haridas, G. 'Tuberculosis in Infants and Children', *Journal of the Malayan Branch of the British Medical Association* 1 (3), December 1937, pp. 238–242.

Haridas, G. 'Tuberculosis among Children Admitted to the Children's Ward, Civil General Hospital, Singapore', *Medical Journal of Malaya* 1 (4), June 1947, pp. 223–237.

Heng, B.H., K.K. Tan; K.W. Chan, and T.H. Tan. 'An Evaluation of 1987 Tuberculosis Deaths in Singapore', *Singapore Medical Journal* 31, 1990, pp. 418–421.

Housing and Development Board. *50,000 Up: Homes for the People* (Singapore: Housing and Development Board, 1966).

Hutchinson, W.E. 'Some Aspects of the Tuberculosis Problem in Singapore', *Journal of the Malayan Branch of the British Medical Association* 1 (3), December 1937, pp. 218–229.

Jarman, Robert L. (ed.). *Annual Reports of the Straits Settlements, 1855–1941* (London: Archive Editions Limited, 1998).

Jones, B.M. 'Summary of Discussion', *Journal of the Malayan Branch of the British Medical Association* 1 (3), December 1937, p. 252.

Lee, Siew Hua. *150 Years of Caring: The Legacy of Tan Tock Seng Hospital* (Singapore: Tan Tock Seng Hospital, 1994).

Lim, Boon Keng. 'Tuberculosis Among the Chinese in Singapore', *Journal of the Malayan Branch of the British Medical Association* January 1904, pp. 16–23.

Lim, Kay Tong and Mary Lee. *Fighting TB: The SATA Story (1947–1997)* (Singapore: Singapore Anti-Tuberculosis Association, 1997).

Lorange, Eric Emik. *Lorange Report: Final Report on Central Redevelopment of Singapore City Prepared for the Govt. of Singapore by Eric Emik Lorange in Capacity of U.N. Town Planning Advisor*. Unpublished report. 1962.

McDougall, Colin. '3-Year Defaulters from a Tuberculosis Out-Patient Clinic', *Medical Journal of Malaya* 9 (2), December 1954, pp. 132–138.

McSwan, D.M. 'A Discussion of Tuberculosis in Malaya', *Medical Journal of Malaya* Vol. IV, 1929, pp. 126–129.

McSwan, D.M. 'The Problem of Tuberculosis with Special Reference to Singapore', *Journal of the Malayan Branch of the British Medical Association,* 1 (3), December 1937, pp. 209–211.

Medical Department. *Annual Reports.*

Ministry of Culture. *Democratic Socialism in Action, June 1959-April 1963* (Singapore: Ministry of Culture, 1963).

Ministry of Health. *Annual Reports*.

Ministry of Health. 'Communicable Diseases Surveillance in 2016'. www.moh.gov.sg/resources-statistics/reports/communicable-diseases-surveillance-in-singapore-2016.

Ministry of Health. Press statement, 'Anti-Tuberculosis Services', 15 March 1975, www.nas.gov.sg/archivesonline/speeches/record-details/7d2538e5-115d-11e3-83d5-0050568939ad.

Ministry of Health. *Singapore 1975 Tuberculosis Prevalence Survey* (Singapore: Ministry of Health, 1978).

National Archives of Singapore. Broadcast by Dr C.E. Smith, Medical Superintendent and chest physician of TTSH, on tuberculosis survey made over Radio Malaya, 5 June 1958, www.nas.gov.sg/archivesonline/speeches/record-details/7615baa6-bcf7-11e6-b045-0050568939ad.

National Archives of Singapore. Text of a Radio Malaya broadcast by Dr C.E. Smith, Ministry of Health, 22 June 1958, www.nas.gov.sg/archivesonline/speeches/record-details/7b3ebe20-bcf7-11e6-b045-0050568939ad.

Pallister, R.A. 'Some Observations on Pulmonary Tuberculosis in Singapore', *Journal of the Malayan Branch of the British Medical Association* 1 (3), December 1937, pp. 231–235.

Pan-Malayan Tuberculosis Conference. *Transactions of the First Pan-Malayan Tuberculosis Conference* 1–4 November 1956 (Singapore: Government Printing Press, 1957).

Proceedings of the Legislative Council of Singapore. *Modifications to the Plan*, February 1948, p. C45.

Proceedings of the Legislative Council of Singapore, *The Medical Plan for Singapore*, 18 May 1948.

Proceedings of the Legislative Council of Singapore. *Tuberculosis Policy: Singapore*, 17 August 1948, p. C212.

Proceedings of the Legislative Council of Singapore. *Report of a Select Committee of the Legislative Council on the Medical Plan for Singapore*, 19 October 1948.

'Report of Dr. A. Morland on Tuberculosis in Malaya', *The Medical Journal of Malaya* 4 (4), June 1950, pp. 272–280.

SATA CommHealth. *The SATA Story: Celebrating 65 Years of Caring for the Community* (Singapore: SATA CommHealth, 2012).

Simpson, W.J. *Report of the Sanitary Condition of Singapore* (London: Waterlow & Sons, 1907).

Singapore. *Proceedings and Report of the Commission Appointed to Inquire into the Cause of the Present Housing Difficulties in Singapore, and the Steps Which Should be Taken to Remedy Such Difficulties*, Vols. I & II (Singapore: Government Printing Office, 1918).

Singapore. *Report of the Housing Committee* (Singapore: Government Printing House, 1947).

Singapore Anti-Tuberculosis Association. *Annual Reports*.

Singapore Anti-Tuberculosis Association. *The Royal Singapore Tuberculosis Clinic of the Singapore Anti-Tuberculosis Association* (Singapore: D. Moore, 1954).

Singapore Government Press Statement. 'Anti-Tuberculosis Week', 8 January 1963, www.nas.gov.sg/archivesonline/speeches/record-details/78be4c28-115d-11e3-83d5-0050568939ad.

Singapore Improvement Trust. *The Work of the Singapore Improvement Trust 1927–1947*.

Singapore Municipality. *Administration Reports*.

Singapore Social Welfare Department, *Annual Reports*.

Speech by Lee Kuan Yew, Secretary-General of the People's Action Party and Prime Minister of Singapore, on 22 September, 1963 after Announcements on the General

Election Results, p. 5, www.nas.gov.sg/archivesonline/speeches/record-details/7409855f-115d-11e3-83d5-0050568939ad.

Speech by Teh Cheang Wan at the URA's 10th anniversary dinner, 30 March 1984, www.nas.gov.sg/archivesonline/speeches/record-details/73544c98-115d-11e3-83d5-0050568939ad.

Speech by the Minister for Health, A.J. Braga, at the Opening of the Institute of Health, 14 May 1958, www.nas.gov.sg/archivesonline/speeches/record-details/dc496573-bcf1-11e6-b045-0050568939ad.

Speech by Yeo Cheow Tong, Minister for Health, at the Launch of the Singapore Tuberculosis Elimination Programme, 4 April 1997, www.nas.gov.sg/archivesonline/speeches/record-details/76487a2b-115d-11e3-83d5-0050568939ad.

Speech of the Minister for Health Yong Nyuk Lin at the Opening of the Anti-Tuberculosis Week Exhibition, 7 April 1964, www.nas.gov.sg/archivesonline/speeches/record-details/78c012cb-115d-11e3-83d5-0050568939ad.

Speech on 'Labour and Welfare Services' by L.C. Goh, Permanent Secretary, Ministry of Labour and Welfare, 23 May 1958, www.nas.gov.sg/archivesonline/speeches/record-details/df4bf655-bcf1-11e6-b045-0050568939ad.

Tan, K.K., A. Cherian and S.K. Teo, 'Tuberculosis in the Elderly', *Singapore Medical Journal* Issue 32, 1991, pp. 423–426.

Teh, Cheang Wan. 'Public Housing in Singapore: An Overview', in Stephen H.K. Yeh (ed.), *Public Housing in Singapore: A Multidisciplinary Study* (Singapore: Singapore University Press for Housing and Development Board, 1975), pp. 1–21.

Treatment Action Group. *Tuberculosis Research Funding Trends 2015–2017.* 2018, www.treatmentactiongroup.org/sites/default/files/tb_funding_2018_final.pdf.

Tuberculosis Control Unit. *Annual Reports.*

Winchester, J.W. 'Observations on Mortality from Tuberculosis in the Straits Settlements', *Medical Journal of Malaya* Vol. IX, 1934, pp. 182–187.

World Health Organization. *The End TB Strategy*, www.who.int/tb/strategy/end-tb/en/.

World Health Organization. *Global Tuberculosis Report 2018*, www.who.int/tb/publications/global_report/en/.

World Health Organization. 'WHO Statement on BCG Revaccination for the Prevention of Tuberculosis', *Bulletin of the World Health Organization* 73 (6), 1995, pp. 805–810.

Written answer by Mr Lim Swee Say, Minister for Manpower, to Parliamentary Question on Unjust Termination of Employment for Employees Diagnosed with Tuberculosis. 2016, www.mom.gov.sg/newsroom/parliament-questions-and-replies/2016/0913-written-answer-by-mr-lim-swee-say-pq-on-unjust-termination-of-employment-for-employees-diagnosed-with-tuberculosis.

Oral history interviews by Oral History Centre, National Archives of Singapore

Abisheganaden, Paul. 17 June 1994.

Burkill, Humphrey Morrison. 1 October 1999.

Chellapah, Wilfred. 25 November 1983.

Chen, Swee Soo. 25 February 2000.

Chew, Andrew. 26 June 1995.

Chew, Benjamin. 27 October 1983.

Chew, Chin Hin. 7 July 1999.

Curtis, Richard John Froude. 17 October 1983.

Gabriel, Vincent. 21 April 2005.
Lee, Liang Hye. 15 April 1985.
Lo, Hong Ling. 22 June 2003.
Loh, Robert Choo Kiat. 6 July 2001.
McNally, Joseph. 8 April 1999.
Monteiro, Edmund Hugh. 16 October 1997.
Mosbergen, Rudy William. 9 June 1994.
Nayar, Cecilia. 25 April 2000.
Oehlers, Farleigh Arthur Charles. 1 June 1984.
Phoon, Winnie. 20 June 2008.
Sen Gupta, N.C. 5 February 1999.
Subramaniom, Bala. 16 October 2008.
Tay, Chin Tian. 3 November 1989.
Toh, Peng Koon. 29 August 1984.

Oral history interviews by Kah Seng Loh

Chee, Cynthia. 2 March 2017.
Chew, Chin Hin. 8 March 2017.
Goh, Kee Tai. 16 February 2017.
Leong, Chew Yin. 30 March 2017.
Pushparani. 14 March 2017.
Tan, Wendy. 28 March 2017.
Tang, Boon H. 13 April 2017.
Teo, Seng Kee, 8 March 2017.

Newspapers

Business Times.
Eastern Daily Mail.
Indian Daily Mail.
Malaya Tribune.
Morning Tribune.
New Nation.
Singapore Free Press.
Singapore Free Press and Mercantile Advertiser.
Singapore Monitor.
Straits Times.
Straits Times Overland Journal.
The New Paper.
Today.

Books and articles

Abel, Emily K. *Tuberculosis and the Politics of Exclusion: A History of Public Health and Migration to Los Angeles* (New Brunswick, NJ: Rutgers University Press, 2007).
Amrith, Sunil S. 'In Search of a Magic Bullet for Tuberculosis: South India and Beyond, 1955–1965', *Social History of Medicine* 17 (1), 2004, pp. 113–130.

Arnold, David (ed.). *Imperial Medicine and Indigenous Societies* (Manchester; New York: Manchester University Press, 1988).

Baker, R.A. and R.A. Bayliss. 'William John Ritchie Simpson (1855–1931): Public Health and Tropical Medicine', *Medical History* 31, 1987, pp. 450–465.

Barr, Michael D. *Singapore: A Modern History* (London: I.B. Tauris, 2019).

Bashford, Alison. *Imperial Hygiene: A Critical History of Colonialism, Nationalism and Public Health* (Houndmills, Basingstoke: Palgrave Macmillan, 2004).

Chan, Heng Chee. *The Dynamics of PAP Dominance: The PAP at the Grassroots* (Singapore: Singapore University Press, 1976).

Chee, C.B., Kyi Win Khin, Jeffery Cutter, and Yee Tang Wang. 'The Imminent Threat of Multidrug-resistant Tuberculosis in Singapore'. *Singapore Medical Journal* 53, 2012, pp. 238–240.

Clancey, Gregory. 'Hygiene in a Landlord State: Health, Cleanliness and Chewing Gum in Late Twentieth Century Singapore'. *Science, Technology and Society* 23 (2), 2018, pp. 214–233.

Das, Sourav, Alex R. Cook, Win Wah, Khin Mar Kyi Win, Cynthia Bin-Eng Chee, Yee Tang Wang and Li Yang Hsu. 'Spatial Dynamics of TB within a Highly Urbanised Asian Metropolis using Point Patterns', *Scientific Reports* 7 (36), 2017, pp. 1–9.

Gamer, Robert E. *The Politics of Urban Development in Singapore* (Ithaca, NY: Cornell University Press, 1972).

Ho, Z.J.M., C.B.E. Chee, R.T. Ong, L.H. Sng, W.L.J. Peh, A.R. Cook, L.Y. Hsu, Y.T. Wang, H.F. Koh, and V.J.M. Lee, 'Investigating a Cluster of Multi-drug Resistant Tuberculosis in a High-rise Apartment Block in Singapore', *International Journal of Infectious Diseases* 67, 2018, pp. 46–51.

Hodge, Joseph Morgan. *Triumph of the Expert: Agrarian Doctrines of Development and the Legacies of British Colonialism* (Athens, OH: Ohio University Press, 2007).

Huff, W.G. 'Entitlements, Destitution and Emigration in the 1930s Singapore Great Depression', *Economic History Review* 54 (3), 2001, pp. 290–323.

Jones, Margaret. 'Tuberculosis, Housing and the Colonial State: Hong Kong, 1900–1950', *Modern Asian Studies* 37, 2003, pp. 653–682.

Jones, Stedman Gareth. *Outcast London: A Study in the Relationship between Classes in Victorian Society* (Oxford: Clarendon Press, 1971).

Kim, Ji Han and Jae-Joon Yim. 'Achievements in and Challenges of Tuberculosis Control in South Korea', *Emerging Infectious Diseases* 21 (11), November 2015, pp. 1913–1920.

Kratoska, Paul H. *The Japanese Occupation of Malaya: A Social and Economic History* (London: C. Hurst, 1998).

Lau, Albert. *A Moment of Anguish: Singapore in Malaysia and the Politics of Disengagement* (Singapore: Times Academic Press, 1998).

Lee, Chien Earn and K. Satku. *Singapore's Health Care System: What 50 Years Have Achieved* (Singapore: World Scientific Publishing, 2015).

Lee, Yong Kiat. *The Medical History of Early Singapore* (Tokyo: Southeast Asian Medical Information Centre, 1978).

Liew, Kai Khiun. 'Myths of Civil Society and its Culture Wars', in Kah Seng Loh, Thum Ping Tjin and Jack Chia (eds.), *Living with Myths in Singapore* (Singapore: Ethos Books, 2017), pp. 203–212.

Loh, Kah Seng. 'Within the Singapore Story: The Use and Narrative of History in Singapore', *Crossroads: An Interdisciplinary Journal of Southeast Asian Studies* 12 (2), 1998, pp. 1–21.

Loh, Kah Seng. *Making and Unmaking the Asylum: Leprosy and Modernity in Singapore and Malaysia* (Petaling Jaya: SIRD, 2009).

Loh, Kah Seng. *Squatters into Citizens: The 1961 Bukit Ho Swee Fire and the Making of Modern Singapore* (NUS Press and Asian Studies Association of Australia, Southeast Asia Series, 2013).

Loh, Kah Seng. 'Emergencities: Experts, Squatters and Crisis in Postwar Southeast Asia', *Asian Journal of Social Science, Special Focus: Reframing Modern and Contemporary Southeast Asia: Transnational Connections, Comparisons, and Mobilities* 44 (6), 2016, pp. 684–710.

Manderson, Lenore. 'Health Services and the Legitimation of the Colonial State: British Malaya 1786–1941', *International Journal of Health Services* 17 (1), 1987, pp. 91–112.

Manderson, Lenore. *Sickness and the State: Health and Illness in Colonial Malaya, 1870–1940* (New York: Cambridge University Press, 1996).

McMillen, Christian W. *Discovering Tuberculosis: A Global History, 1900 to the Present* (New Haven, CT and London: Yale University Press, 2015).

O'Connor, Philip. *Britain in the Sixties: Vagrancy* (London: Penguin Books, 1963).

Ryymin, Teemu. 'Civilizing the "Uncivilized": The Fight against Tuberculosis in Northern Norway at the Beginning of the Twentieth Century', *Acta Borealia: A Nordic Journal of Circumpolar Societies* 24 (2), 2007, pp. 143–161.

Saw Swee Hock. *The Population of Singapore* (Singapore: Institute of Southeast Asian Studies, 1999).

Trocki, Carl A. *Opium and Empire: Chinese Society in Colonial Singapore, 1800–1910* (Ithaca, NY: Cornell University Press, 1990).

Turnbull, C.M. *A History of Modern Singapore, 1819–2005* (Singapore: NUS Press, 2009).

Warren, James Francis. *Rickshaw Coolie: A People's History* (Singapore: Singapore University Press, 2003).

Welshman, John. 'Tuberculosis, "Race", and Migration, 1950–70', *Sociology of Health and Illness* 22 (6), November 2000, pp. 858–882.

Wong, Lin Ken. 'The Trade of Singapore, 1819–69'. *Journal of the Malayan Branch of the Royal Asiatic Society* 33, Part 4 (192), 1961, pp. 4–315.

Worboys, Michael 'Tuberculosis and Race in Britain and its Empire', in Ernst Waltraud and Bernard Harris (eds.), *Race, Science and Medicine, 1700–1960* (London: Routledge, 1999), pp. 144–166.

Yap, P., K.H.X. Tan, W.Y. Lim, T. Barkham, L.W.L. Tan, M.I. Chen, Y.T. Wang, and C.B.E. Chee, 'Prevalence of and Risk Factors Associated with Latent Tuberculosis in Singapore: A Cross-sectional Survey'. *International Journal of Infectious Diseases* 72, July 2018, pp. 55–62.

Yeo, Kim Wah. *Political Development in Singapore, 1945–55* (Singapore: Singapore University Press, 1973).

Yeoh, Brenda S.A. *Contesting Space in Colonial Singapore: Power Relations and the Urban Built Environment* (Singapore: Singapore University Press, 2003), 2nd edition.

Index